Springer Theses

Recognizing Outstanding Ph.D. Research

Aims and Scope

The series "Springer Theses" brings together a selection of the very best Ph.D. theses from around the world and across the physical sciences. Nominated and endorsed by two recognized specialists, each published volume has been selected for its scientific excellence and the high impact of its contents for the pertinent field of research. For greater accessibility to non-specialists, the published versions include an extended introduction, as well as a foreword by the student's supervisor explaining the special relevance of the work for the field. As a whole, the series will provide a valuable resource both for newcomers to the research fields described, and for other scientists seeking detailed background information on special questions. Finally, it provides an accredited documentation of the valuable contributions made by today's younger generation of scientists.

Theses are accepted into the series by invited nomination only and must fulfill all of the following criteria

- They must be written in good English.
- The topic should fall within the confines of Chemistry, Physics, Earth Sciences, Engineering and related interdisciplinary fields such as Materials, Nanoscience, Chemical Engineering, Complex Systems and Biophysics.
- The work reported in the thesis must represent a significant scientific advance.
- If the thesis includes previously published material, permission to reproduce this must be gained from the respective copyright holder.
- They must have been examined and passed during the 12 months prior to nomination.
- Each thesis should include a foreword by the supervisor outlining the significance of its content.
- The theses should have a clearly defined structure including an introduction accessible to scientists not expert in that particular field.

More information about this series at http://www.springer.com/series/8790

Aidan Strathearn

Modelling Non-Markovian Quantum Systems Using Tensor Networks

Doctoral Thesis accepted by
University of St Andrews, Fife, United Kingdom

Springer

Author
Dr. Aidan Strathearn
School of Mathematics and Physics
University of Queensland
Brisbane, QLD, Australia

Supervisor
Prof. Brendon Lovett
School of Physics and Astronomy
University of St Andrews
Fife, UK

ISSN 2190-5053 ISSN 2190-5061 (electronic)
Springer Theses
ISBN 978-3-030-54974-9 ISBN 978-3-030-54975-6 (eBook)
https://doi.org/10.1007/978-3-030-54975-6

This Springer imprint is published by the registered company Springer Nature Switzerland AG
The registered company address is: Gewerbestrasse 11, 6330 Cham, Switzerland

For Charles,
my cat,
my dead cat.

Supervisor's Foreword

Quantum mechanics has been a pillar of theoretical physics for over a century, and remains the key to understanding the world of atoms and molecules that surrounds us. For example, the emission spectrum of a hydrogen atom can be well understood by solving the Schrödinger equation for an electron in the electric field produced by a proton.

However, as the number of particles in the system goes up, it becomes increasingly difficult to solve the quantum mechanical equations and predict behaviour. This is because the number of possible quantum states scales exponentially with the number of particles, and so even for a few tens of interacting particles an exact quantum description becomes impossible.

This presents us with a problem. Many nanoscale devices that are now routinely fabricated in the laboratory must behave quantum mechanically, but they contain many billions of atoms. How can we then predict and engineer their behaviour when we cannot solve the underlying quantum mechanical equations? The solution of this problem lies at the heart of this thesis 'Modelling Non-Markovian Quantum Systems Using Tensor Networks'.

The key observation which allows us to make progress with such problems is that a nanodevice often consists of just a few key 'system' quantum particles whose behaviour we can control and measure, and so the problem of modelling them becomes a question of how these few particles behave in the 'environment' of all the others. In particular, this picture allows us to make some assumptions and approximations that enable us to start to tackle the problem.

A important simplification is made by assuming that the environment consists of a set of harmonic oscillators, and that before interacting with the system these are all in thermal equilibrium. This approximation can apply very widely, since to lowest order the bottom of a potential well—which particles will be close to at low enough temperature—looks parabolic. With this approximation in place, the state of the environment is *Gaussian* and requires much less information to describe it than a general state.

This assumption was enough for Richard Feynman to expand his path integral picture of how a quantum system evolves to include the environment. In fact, he found that the only effect of the environment is to introduce extra phase shifts in the paths in phase space of the system, and that these shifts—the so-called *influence functional*—can be found *exactly*. This was a huge step forward since now we had a way of describing just a few 'system' particles in the presence of a harmonic environment. When Nancy Makri showed how to discretize time in this description so that the dynamics could be simulated on a computer, open quantum systems could be accurately simulated. However, this could only be done for a few time steps, since the possible paths in Feynman's description divide at every step. We are faced with an exponentially growing number of paths which quickly take up all the available memory on any computer.

My colleague Dr. Jonathan Keeling had spoken to me around 2012 about the possibility of recasting Feynman's method as a tensor network. Tensor networks represent ways of decomposing larger tensors into smaller elements, and had been successfully used to describe the behaviour of several quantum particles interacting in space. Jonathan wondered if we might do something similar for the behaviour of a few particles in an environment, using Feynman's method. Crucially, though, the input of a very talented young theoretical physicist was needed to actually get this to work. In 2014 I met Aidan Strathearn; in fact he popped over to see me on a Sunday morning since, extremely fortuitously, his family lives opposite. He was inspired by the idea of the project, and the final result of his Ph.D. studies appear in this book. In it, he reveals his deep insights into how open quantum dynamics can be captured by building and contracting tensor networks. His work is truly remarkable: he develops a whole new way of thinking about how to represent an influence functional. He also shows how efficient it is: we replace exponentially growing complexity with a polynomial scaling, allowing the simulation of many hundreds of time steps for a single system particle. He applies the technique to several examples, showing how accurate and versatile its results are.

This is a very important piece of work that has great potential for future development and research into the quantum systems that will underpin much of the technology of the twenty-first century.

May 2020

Prof. Brendon Lovett
School of Physics and Astronomy
University of St Andrews
Fife, UK

Abstract

Accurately modelling the behaviour of quantum systems interacting with the environment is vital to the development of quantum technology. Open quantum systems are well understood when the influence of their environment is weak, but the problem of modelling general systems away from this limit remains difficult. Currently, all approaches to making this problem tractable require some assumption about the nature of the system and its environment. In this thesis a general and efficient numerical method for calculating observables of open quantum systems is presented. The method is to express the exact equations of motion that describe the evolution of a general open quantum system as a tensor network, whose structure allows for a decomposition in terms of matrix products that can be efficiently compressed in size. The power and versatility of the method is demonstrated by using it to study three contrasting models. The first of these is the Ohmic Spin-Boson model, in which the location of the localisation phase transition is identified by analysing the dissipative spin dynamics. This requires high precision numerical calculations, which are shown to be carried out with high efficiency. The second model is that of two spatially separated spins interacting with the same environment, for which no other exact results are available. Here the environment is seen to mediate interaction between the spins, which is intuitively found to be longer ranged for a lower spatial dimension. Finally, an experimentally relevant model of a driven quantum dot is considered. Calculations of the emission spectrum of the dot reveal complex interplay between the coherent driving and the dissipative influence of phonons. In particular the phonon sideband in the spectrum is found to be supressed as the driving of the dot is strengthened.

Acknowledgements

From the day I came knocking on the door of his home to enquire about doing a Ph.D. to the day I came knocking to bid him farewell as I departed for my first academic job, my supervisor Brendon Lovett has shown me kindness, support, and encouragment. It has been a pleasure working with him so far and I am deeply grateful for his supervision. I would also like to thank Peter Kirton. From my experience the most rich and stimulating scientific discussions are the ones that get heated, of which Peter and I have had many. I hope there are many more to come. I am also delighted to have had the opportunity to work with Jonathan Keeling who has shown me how a true pro does science. I shall not name everyone who has enriched my experience doing a Ph.D. in St. Andrews but I will say a big thank you to Dominic Gribben, who I have had great fun collaborating with, Kyle Ballantine, who I have had great fun drinking with, and Matt Trott, who I have had great fun living with. Both Brendon and Peter were responsible for proofreading this thesis, so I extend my thanks to them both especially for this.

As I write this I am about 10,000 miles separated from my family and missing them very much. My parents, Sally and Sandy, have always given me their love and support, without which I could never made it this far. My sister, Rebecca, has always called me names and burped in my face, but I still love her.

Finally, a big up to all the homeboys: Martin "YungMertz" Davidson, Oliver "NikNak" Gardner, Douglas "nugget92" Braithwaite, James "House" Houston, James "Hound" Barker, Russell "Swagmaster" Hope, Kyle "Lafferty Lawdd" Lafferty, and Neil "Haigy" Haig.

Acknowledgements

Contents

Abbreviations

ADT	Augmented density tensor
HB	Harmonic bath
IF	Influence functional
MPO	Matrix product operator
MPS	Matrix product state
QRT	Quantum regression theorem
QUAPI	Quasi-adiabatic propagator path integral
RDM	Reduced density matrix
RSM	Redfield-Secular-Markov
SBM	Spin-boson model
SD	Spectral density
SVD	Singular value decomposition
TEMPO	Time-evolving matrix product operator
TLS	Two-level system
TN	Tensor network

Chapter 1
Introduction

The ever improving capability to investigate and control systems at the atomic-scale, such as trapped ions [1] and quantum dots [2], means that understanding the quantum mechanical nature of such systems is vital. These systems must be considered as open quantum systems, since any physical system addressed in experiment cannot be considered to be truly isolated from its environment. Gaining an understanding of the behaviour of open quantum systems is important both in the development of new quantum technology, where scalability and reliability hinge upon a robust understanding of sources of noise, and in performing stringent checks of the mathematical models upon which this understanding can be built. For example, the realisation of large scale quantum computation is coming ever closer to reality and ongoing development relies crucially on a solid theoretical framework to model open quantum systems [3–6]. The design of efficient photo-voltaic devices is another endeavor with the potential for profound impact on society and for which an understanding of the interplay between system and environment is key [7–11].

The theory of open quantum systems [12–14] was initially developed to model atoms emitting and absorbing optical radiation, whose behaviour can be described by constant rates of decay and excitation of atomic energy levels [15, 16]. An open quantum system that is well described by a constant rate equation is said to be Markovian [13], meaning its behaviour at a given time is independent of its behaviour in the past. Markovian quantum systems commonly occur in nature [17], usually when coupling to the environment is weak and unstructured, so methods of modelling them are well refined. The general theoretical framework used to model Markovian quantum systems is mathematically rigorous and has proven to be highly effective in describing observations made in experiment [18]. The fundamental equations of motion that govern the dynamics of Markovian quantum systems generalise the well known Schrödinger equation, providing a means to model thermalisation and decoherence, and their general allowed form has been thoroughly established [19, 20].

A. Strathearn, *Modelling Non-Markovian Quantum Systems Using Tensor Networks*, Springer Theses, https://doi.org/10.1007/978-3-030-54975-6_1

Constant rate equations that describe Markovian quantum systems cannot account for the fact that the environment of an open system will be dynamically perturbed by its behaviour. Changes in the environment induced by the system at one instance in time will affect how the environment influences the system at a later time. A Markovian system is one for which this backaction upon the system due to environment can be ignored, but when such effects must be accounted for to effectively describe the behaviour of an open system it is referred to as non-Markovian [21, 22]. Strictly speaking then, since the backaction of the environment upon the system is never truly absent unless they decouple, all open systems are non-Markovian to some degree. To give a physical example, the periodic exchange of a single quantum of energy between an atom (the system) and an electromagnetic field (the environment) would qualify the atom as non-Markovian. The general allowed form of the equations of motion for a non-Markovian quantum system is not known and currently a practical theoretical quantification of non-Markovianity in quantum systems has yet to be agreed upon [23]. Despite the relatively underdeveloped state of the theory of non-Markovian quantum systems compared to that of Markovian quantum systems, significant progress is being made through both experimental and theoretical work in the field. For instance, clear non-Markovian behaviour has been observed in micromechanical oscillators [24] and superconducting circuits [25], while theoretical guidance has led to the engineering and control of non-Markovianity in photonic systems [26–28]. Gaining experimental control over non-Markovian behaviour not only helps establish the underlying theory, but potentially allows for the use of non-Markovianity as a resource for improving the operation of quantum devices [29–31].

Tensor network theory has had transformative impact on the way quantum many-body systems are studied [32–34]. Typically the state of a quantum system remains confined only to subregions of the full state space it inhabits [32], and by representing such systems as connected series of tensors, tensor networks, the components that are irrelevant to the effective description the system can be identified and discarded, significantly reducing the size of the problem. Notably, tensor networks provide highly efficient methods to calculate both dynamics [35–37] and equilibrium properties [38–40] of one-dimensional systems. Representing problems as tensor networks not only allows for their efficient solution, but can also lend insight their nature. The structure of a tensor network representing a quantum system is often intricately linked with the entanglement that is present [32], for example. Approaches to modelling non-Markovian quantum systems have also been developed using tensor networks, by treating open few-body systems and their environments collectively as closed many-body systems [41, 42]. Such approaches are the state-of-the-art in the field, providing exact simulations that have shed light on the role of non-Markovianity in realistic systems [43].

In this thesis a new method for modelling non-Markovian quantum systems using tensor networks is introduced and implemented with high efficiency to study a variety of open systems in an exact way. Representation of problems as tensor networks is achieved in a novel way, essentially by treating instances of a single open system at many points in time as a many-body problem. This approach is not derived from properties of the system but rather by identifying that any problem involving Dyson time ordering [44] has a tensor network representation. Using this result to write the exact equations that describe open quantum systems in tensor network form then makes them solvable using standard techniques of tensor network theory. This new tensor network approach to time ordering problems has connections to the field of Feynman operator calculus [45–47], where it can be thought of as a disentangling technique, and to the process tensor description of quantum systems [48, 49], where it enables a tensor network decomposition of the process tensor for a given open system.

The layout of this thesis is as follows. In Chap. 2 the established theory of open quantum systems, both Markovian and non-Markovian, is reviewed and the relevant concepts in tensor network theory required for the working of the new methods to be presented are introduced. In Chap. 3 a method for solving arbitrary models of open quantum systems using tensor networks is proposed and a proof-of-principle demonstration of the method is given. In Chap. 4 the methods developed in Chap. 3 are applied to investigate the physics of three contrasting open quantum systems: a dissipative quantum phase transition is identified in the Spin-Boson model, non-Markovian environment mediated interactions are found to occur between two spatially separated spins, and nontrivial features in the emission spectrum of a driven quantum dot are uncovered. Chapter 5 is a summary, which includes a discussion on ongoing applications of the methodology introduced here and an outlook on how it can be more broadly applied.

Publications and Contributions

Part of the work of this thesis has led to the following publications:

Publication: A. Strathearn, B. W. Lovett, and P. Kirton. Efficient real-time path integrals for non-Markovian spin-boson models. *New Journal of Physics*, 19(9):093009, 2017.
Contribution: Conception and implementation of main theoretical result, all application and data analysis, writing of the initial draft of the paper and contributions to subsequent editing.

Publication: A. Strathearn, P. Kirton, D. Kilda, J. Keeling, and B. W. Lovett. Efficient non-Markovian quantum dynamics using time-evolving matrix product operators, *Nature Communications*. 9:3322, 2018.
Contribution: Lead development and implementation of main result, major contribution to application and data analysis, contributions at all stages of writing process.

References

1. Lv D, An S, Liu Z, Zhang J-N, Pedernales JS, Lamata L, Solano E, Kim K (2018) Quantum simulation of the quantum Rabi model in a trapped ion. Phys Rev X 8(2):021027
2. Ramsay AJ, Gopal AV, Gauger EM, Nazir A, Lovett BW, Fox AM, Skolnick MS (2010) Damping of exciton Rabi rotations by acoustic phonons in optically excited InGaAs/GaAs quantum dots. Phys Rev Lett 104:017402
3. Preskill J (2018) Quantum computing in the NISQ era and beyond. Quantum 2:79
4. DiVincenzo D, Terhal B (1998) Decoherence: the obstacle to quantum computation. Phys World 11(3):53
5. Devitt SJ, Munro WJ, Nemoto K (2013) Quantum error correction for beginners. Rep Prog Phys 76(7):076001
6. Chirolli L, Burkard G (2008) Decoherence in solid-state qubits. Adv Phys 57(3):225
7. Semonin OE, Luther JM, Beard MC (2012) Quantum dots for next-generation photovoltaics. Mater Today 15(11):508
8. Yang Z, Fan JZ, Proppe AH, García de Arquer FP, Rossouw D, Voznyy O, Lan X, Liu M, Walters G, Quintero-Bermudez R, Sun B, Hoogland S, Botton GA, Kelley SO, Sargent EH (2017) Mixed-quantum-dot solar cells. Nat Commun 8:1325
9. Nozik AJ (2002) Quantum dot solar cells. Phys E: Low-dimens Syst Nanostruct 14(1–2):115
10. Uchiyama C, Munro W J, Nemoto K (2018) Environmental engineering for quantum energy transport. npj Quantum Inf 4:33
11. Rebentrost P, Mohseni M, Kassal I, Lloyd S, Aspuru-Guzik A (2009) Environment-assisted quantum transport. New J Phys 11(3):033003
12. Rivas Á, Huelga SF (2012) Open Quantum Systems. Springer, Berlin
13. Breuer H-P, Petruccione F (2002) The Theory of Open Quantum Systems. Oxford University Press, Oxford
14. Weiss U (2012) Quantum Dissipative Systems. World Scientific
15. Carmichael HJ (1999) Statistical Methods in Quantum Optics 1. Springer, Berlin
16. Walls DF, Milburn GJ (2007) Quantum Optics, 2nd edn. Springer, Berlin
17. Figueroa-Romero P, Modi K, Pollock FA (2019) Almost Markovian processes from closed dynamics. Quantum 3:136
18. Gardiner C, Zoller P (2004) Quantum Noise: A Handbook of Markovian and Non-Markovian Quantum Stochastic Methods with Applications to Quantum Optics. Springer Series in Synergetics. Springer, Berlin
19. Lindblad G (1976) On the generators of quantum dynamical semigroups. Commun Math Phys 48(2):119
20. Gorini V (1976) Completely positive dynamical semigroups of n-level systems. J Math Phys 17(5):821
21. de Vega I, Alonso D (2017) Dynamics of non-Markovian open quantum systems. Rev Mod Phys 89:015001
22. Breuer H-P, Laine E-M, Piilo J, Vacchini B (2016) Colloquium: non-Markovian dynamics in open quantum systems. Rev Mod Phys 88:021002
23. Rivas Á, Huelga SF, Plenio MB (2014) Quantum non-Markovianity: characterization, quantification and detection. Rep Prog Phys 77(9):094001
24. Gröblacher S, Trubarov A, Prigge N, Cole GD, Aspelmeyer M, Eisert J (2015) Observation of non-Markovian micromechanical Brownian motion. Nat Commun 6:7606
25. Potočnik A, Bargerbos A, Schröder FAYN, Khan SA, Collodo MC, Gasparinetti S, Salathé Y, Creatore C, Eichler C, Türeci HE, Chin AW, Wallraff A (2018) Studying light-harvesting models with superconducting circuits. Nat Commun 9:904
26. Liu B-H, Cao D-Y, Huang Y-F, Li C-F, Guo G-C, Laine E-M, Breuer H-P, Piilo J (2013) Photonic realization of nonlocal memory effects and non-Markovian quantum probes. Sci Rep 3:1781

27. Liu B-H, Li L, Huang Y-F, Li C-F, Guo G-C, Laine E-M, Breuer H-P, Piilo J (2011) Experimental control of the transition from Markovian to non-Markovian dynamics of open quantum systems. Nat Phys 7:931
28. Liu Z-D, Lyyra H, Sun Y-N, Liu B-H, Li C-F, Guo G-C, Maniscalco S, Piilo J (2018) Experimental implementation of fully controlled dephasing dynamics and synthetic spectral densities. Nat Commun 9:3453
29. Thomas G, Siddharth N, Banerjee S, Ghosh S (2018) Thermodynamics of non-Markovian reservoirs and heat engines. Phys Rev E 97(6):062108
30. Bylicka B, Chruściński D, Maniscalco S (2014) Non-Markovianity and reservoir memory of quantum channels: a quantum information theory perspective. Sci Rep 4:5720
31. Sakuldee F, Milz S, Pollock FA, Modi K (2018) Non-Markovian quantum control as coherent stochastic trajectories. J Phys A: Math Theor 51(41):414014
32. Orús R (2014) A practical introduction to tensor networks: matrix product states and projected entangled pair states. Ann Phys (N.Y.) 349(Supplement C):117
33. Verstraete F, Murg V, Cirac JI (2008) Matrix product states, projected entangled pair states, and variational renormalization group methods for quantum spin systems. Adv Phys 57(2):143–224
34. Schuch N, Pérez-García D, Cirac I (2011) Classifying quantum phases using matrix product states and projected entangled pair states. Phys Rev B 84(16):165139
35. Cazalilla MA, Marston JB (2002) Time-dependent density-matrix renormalization group: a systematic method for the study of quantum many-body out-of-equilibrium systems. Phys Rev Lett 88(25):256403
36. Daley AJ, Kollath C, Schollwöck U, Vidal G (2004) Time-dependent density-matrix renormalization-group using adaptive effective Hilbert spaces. J Stat Mech Theory Exp 2004(04):P04005
37. Vidal G (2004) Efficient simulation of one-dimensional quantum many-body systems. Phys Rev Lett 93(4):040502
38. White SR (1992) Density matrix formulation for quantum renormalization groups. Phys Rev Lett 69(19):2863
39. Verstraete F, Cirac I (2006) Matrix product states represent ground states faithfully. Phys Rev B 73(9):094423
40. Schollwöck U (2005) The density-matrix renormalization group. Rev Mod Phys 77(1):259
41. Chin AW, Rivas Á, Huelga SF, Plenio MB (2010) Exact mapping between system-reservoir quantum models and semi-infinite discrete chains using orthogonal polynomials. J Math Phys 51(9):092109
42. Prior J, Chin AW, Huelga SF, Plenio MB (2010) Efficient simulation of strong system-environment interactions. Phys Rev Lett 105(5):050404
43. Chin AW, Prior J, Rosenbach R, Caycedo-Soler F, Huelga SF, Plenio MB (2013) The role of non-equilibrium vibrational structures in electronic coherence and recoherence in pigment–protein complexes. Nat Phys 9(2):113
44. Dyson FJ (1949) The radiation theories of Tomonaga, Schwinger, and Feynman. Phys Rev 75(3):486
45. Feynman RP (1951) An operator calculus having applications in quantum electrodynamics. Phys Rev 84(1):108
46. Jefferies B, Johnson GW, Kim BS (2006) Feynman's operational calculi: methods for iterative disentangling. Acta Applicandae Mathematicae 92(3):293–309
47. Gill TL (2017) The Feynman-Dyson view. J Phys Conf Ser 845:012023
48. Pollock FA, Rodríguez-Rosario C, Frauenheim T, Paternostro M, Modi K (2018) Non-Markovian quantum processes: complete framework and efficient characterization. Phys Rev A 97:012127
49. Jørgensen MR, Pollock FA (2019) Exploiting the causal tensor network structure of quantum processes to efficiently simulate non-Markovian path integrals. Phys Rev Lett 123:240602

Chapter 2
Background

2.1 Theory of Open Quantum Systems

In this section the elementary concepts used to construct and solve models for open quantum systems are reviewed and demonstrated. First, a general overview of the techniques required to allow for a quantum mechanical description of open systems is presented in Sect. 2.1.1. A key concept here is that one must always begin with a quantum mechanical model of a closed system, consisting of the system, which is considered open, and the environment to which it couples. With such a model at hand, it is possible to exactly eliminate the degrees of freedom of the environment to, at least formally, obtain a model for the degrees of freedom of the open system alone, detailed in Sect. 2.1.1.1. Such models are described by a time-nonlocal Liouvillian, and perturbation theory can be employed to find a general but simple closed form for this Liouvillian, the subject of Sect. 2.1.1.2. A specific model for an environment, the harmonic bath model, is discussed in Sect. 2.1.1.3. This model is of special importance since an exact closed form Liouvillian can be found to model its influence on an open system. The archetypal model for an open system, the two-level system, and its relevance in practical applications is then reviewed in Sect. 2.1.1.4.

Even with a closed form model of an open system at hand, the time-nonlocality present in such models generally makes them intractable. For many systems it is known that effects of time-nonlocality are negligible. Such systems can loosely be classified as Markovian systems, and the standard techniques that are used to find effective time-local models for these systems are presented in Sect. 2.1.2. First, the general allowed form of a time-local Liouvillian is stated in Sect. 2.1.2.1. In Sect. 2.1.2.2 the primary approximations that are made upon time-nonlocal Liouvillians to put them into this form, the Redfield approximations, are reviewed. The well known Born-Markov approximation and its connection to the Redfield approximations are then given in Sect. 2.1.2.3. Finally, in Sect. 2.1.2.4 the concepts just described are applied and demonstrated using the two-level system as an example.

A. Strathearn, *Modelling Non-Markovian Quantum Systems Using Tensor Networks*, Springer Theses, https://doi.org/10.1007/978-3-030-54975-6_2

Systems in which time-nonlocality must be accounted for are non-Markovian, and for these systems there are no techniques available which enable their solution generally. The content of Sect. 2.1.3 is therefore a review of a variety of techniques developed to tackle specific kinds of non-Markovian system. Some instances of exactly solvable non-Markovian systems, and what can be learned from such systems, are discussed in Sect. 2.1.3.1. The discussion in Sect. 2.1.3.2 is on techniques that involve unitarily transforming into a new frame in which a non-Markovian system can appear to be Markovian. The methods discussed in Sect. 2.1.3.3 are similar in spirit, but rather involve extending the boundary of a non-Markovian system to make it a Markovian system. Another approach to studying non-Markovian systems is described in Sect. 2.1.3.4, involving the introduction of fictitious degrees into such a system with which non-Markovianity can be modelled explicitly.

2.1.1 General Theory

2.1.1.1 Eliminating the Environment

The state of a quantum system at any given time, t, is represented by a density matrix, $\rho(t)$, which is a statistical distribution of pure quantum states. The eigenvalues of a density matrix are the probabilities of finding the system in various orthogonal states, and so must sum to one, be real, and be positive. Their summing to one requires $\mathrm{Tr}\,\rho(t) = 1$, where Tr denotes the trace. Hermiticity, $\rho(t) = \rho^\dagger(t)$, is required to ensure that the eigenvalues of $\rho(t)$ are real, and their positivity requires $\rho(t)$ be a positive semi-definite matrix.

Typically, the system is partitioned into an experimentally addressable part, A, and an uncontrolled environment, B, to which it couples, visualised in Fig. 2.1a. The unitary evolution of the system can be modelled with the Hamiltonian

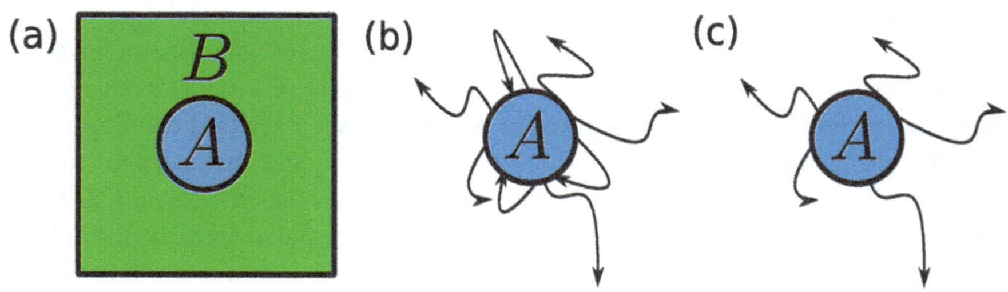

Fig. 2.1 **a** A schematic of a generic closed system consisting of an experimentally addressable subsystem, A, immersed in an uncontrollable environment, B. **b** A schematic of a generic non-Markovian open system, A. Arrowed lines indicate implicit transfer of excitations from A to B that may return to A. **c** A schematic of a generic Markovian open system, A. Arrowed lines again indicate implicit excitation transfer from A to B, but now these excitations leave A never to return

$$H = H_0 + V. \tag{2.1}$$

The noninteracting part, H_0, is decomposed as

$$H_0 = H_A + H_B, \tag{2.2}$$

where H_A and H_B are the Hamiltonians for A and B respectively, while the most general form of interaction potential is [1]

$$V = \sum_{j=1}^{d^2} x_j F_j, \tag{2.3}$$

for some sets of operators $x = \{x_j\}$ and $F = \{F_j\}$ of A and B respectively, with the possibility $[x_j F_j, x_{j'} F_{j'}] \neq 0$. It has been assumed that A is finite with d possible orthogonal states, while B must strictly be infinite and continuous to have a genuinely irreversible influence on A. For notational brevity, this interaction shall be written simply as

$$V = xF, \tag{2.4}$$

which should be understood as an inner product between the vectors of operators x and F. It is assumed that it can always be arranged such that $x_j = x_j^\dagger$. In the special case that $[x_j, x_{j'}] = 0$ for all j and j' it is possible to simultaneously diagonalise all x_j in which case the interaction shall be called abelian.

The density matrix, $\rho(t)$, of the system satisfies the equation [1]

$$\frac{\partial \rho(t)}{\partial t} = -i[H, \rho(t)], \tag{2.5}$$

where the reduced Planck constant has been set to one, $\hbar = 1$. It will be convenient to work in the interaction frame by performing a unitary transformation on all operators of A and B

$$O \rightarrow e^{iH_A t} e^{iH_B t} O e^{-iH_B t} e^{-iH_A t}. \tag{2.6}$$

Then the transformed density matrix satisfies the time dependent equation

$$\frac{\partial \rho(t)}{\partial t} = -i[V(t), \rho(t)], \tag{2.7}$$

where

$$V(t) = x(t)F(t), \tag{2.8}$$

with the time dependence of $x(t) = \{x_j(t)\}$ and $F(t) = \{F_j(t)\}$ as given by Eq. (2.6).

It is assumed that at some instant in time, $t = 0$, A and B are uncorrelated, such that the density matrix factorises,

$$\rho(0) = \rho_A(0)\rho_B(0), \tag{2.9}$$

where $\rho_A(0)$ is the density matrix for A and $\rho_B(0)$ is the density matrix for B. Given this boundary condition, the solution of Eq. (2.7) can formally be expressed in terms of a propagator superoperator [1]

$$\Lambda(t_b, t_a) = \overleftarrow{\mathrm{T}}\, e^{\int_{t_a}^{t_b} \mathcal{L}(t')dt'}, \tag{2.10}$$

where $\overleftarrow{\mathrm{T}}$ denotes time ordering of superoperators from right to left, and the Liouvillian superoperator is defined as

$$\mathcal{L}(t)\rho = -i[V(t), \rho]. \tag{2.11}$$

such that

$$\rho(t) = \Lambda(t, 0)\rho(0). \tag{2.12}$$

The observable quantities of the system are determined by its correlation functions. For example

$$\langle X(t_2)X(t_1)\rangle = \mathrm{Tr}[\,\overleftarrow{\mathrm{T}}\, X(t_2)X(t_1)e^{\int_0^{t_2} \mathcal{L}(t')dt'}\rho(0)], \tag{2.13}$$

which, assuming $t_2 > t_1 > 0$, can be rearranged in time order such that it can also be written in terms of the propagator

$$\langle X(t_2)X(t_1)\rangle = \mathrm{Tr}[X(t_2)\Lambda(t_2, t_1)X(t_1)\Lambda(t_1, 0)\rho(0)]. \tag{2.14}$$

It is often not possible to model the environment, B, explicitly. The standard procedure to eliminate the degrees of freedom of B is to perform a partial trace over expressions used to calculate observables of A [1, 2]. For example, a reduced density matrix (RDM) for A can be found formally by partially tracing over Eq. (2.12),

$$\rho_A(t) = \mathrm{Tr}_B[\rho(t)] = \Lambda_A(t, 0)\rho_A(0), \tag{2.15}$$

where

$$\Lambda_A(t, 0) \equiv \mathrm{Tr}_B\left[\overleftarrow{\mathrm{T}}\, e^{\int_0^{t} \mathcal{L}(t')dt'}\rho_B(0)\right], \tag{2.16}$$

is the reduced propagator and Tr_B denotes the partial trace over B. In certain situations [3, 4] it is possible to obtain a memory kernel master equation for the RDM,

$$\frac{\partial \rho_A(t)}{\partial t} = \int_0^t dt' K(t, t')\rho_A(t'), \tag{2.17}$$

where $K(t, t')$ is the memory kernel superoperator. This master equation describes the behaviour of the RDM at a given time t as depending on the entire previous evolution of the RDM at times $t' \in [0, t]$. The physical interpretation is that A perturbs

B at a time t', creating excitations in B which can then return at a later time t, see Fig. 2.1b. Since the degrees of freedom of B have been traced out, this mechanism is implicit and manifests mathematically as time-nonlocality of the memory kernel.

Calculation of the RDM of A as in Eq. (2.15) provides useful but incomplete information about the behaviour of A. This is because knowledge of the RDM and the reduced propagator only provides enough information to find single time expected observables of the system. That is, despite the suggestive form of Eq. (2.14), it is generally not possible to express multitime correlations in terms of the reduced propagator [5],

$$\langle X(t_2)X(t_1)\rangle \neq \mathrm{Tr}_A[X(t_2)\Lambda_A(t_2,t_1)X(t_1)\Lambda_A(t_1,0)\rho_A(0)]. \tag{2.18}$$

There are some cases where this inequality can be made, exactly or approximately, into an equality, in which case the expression of correlation functions in terms of the reduced propagator is called the quantum regression theorem (QRT) [1, 6]. The situations in which the QRT holds will be discussed further throughout this section.

An object that can be used to model the general observables of an open system can be found by noting that the combination of taking a trace and time ordering, i.e. the operation $\mathrm{Tr}\,\overleftarrow{\mathrm{T}}$, amounts to functional integration over the degrees of freedom of both A and B, $x(t)$ and $F(t)$ respectively. Calculating general observables via the formal operation $\mathrm{Tr}\,\overleftarrow{\mathrm{T}}$, as in Eq. (2.13), can then be carried out by performing two functional integrals, one over $x(t)$ and one over $F(t)$. Thus it is possible to carry out the integral over $F(t)$ independently of $x(t)$, achieved by making the decomposition [1, 7]

$$\mathrm{Tr}\,\overleftarrow{\mathrm{T}} = \left(\mathrm{Tr}_A\,\overleftarrow{\mathrm{T}}_A\right)\left(\mathrm{Tr}_B\,\overleftarrow{\mathrm{T}}_B\right), \tag{2.19}$$

where the subscript A (B) indicates operations on degrees of freedom of A (B) only.

The partitioning of time ordering, $\overleftarrow{\mathrm{T}} = \overleftarrow{\mathrm{T}}_A\,\overleftarrow{\mathrm{T}}_B$, used in this decomposition is valid provided that operators of A and B commute, since for some operator, $O(t) = O_A(t)O_B(t)$, and times, $t_2 > t_1$, it implies

$$O(t_2)O(t_1) = O_A(t_2)O_B(t_2)O_A(t_1)O_B(t_1) = O_A(t_2)O_A(t_1)O_B(t_2)O_B(t_1) \tag{2.20}$$

where $O_A(t)$ and $O_B(t)$ are operators of A and B, respectively. The first equality shows the result of arranging the O operators in time order while the second equality shows the result of arranging the O_A and O_B operators in time order separately. Commutativity between operators of A and B is guaranteed trivially if they are bosonic or spin systems, making the decomposition, Eq. (2.19), generally possible for such systems. If A and B are fermionic then the T_A and T_B appearing can each be replaced by the full time ordering operator, $\overleftarrow{\mathrm{T}}$, which incurs only minor changes to the arguments presented in the following sections. This replacement is possible independent of the nature of A and B since time ordering is an idempotent operation, i.e. $\overleftarrow{\mathrm{T}} = \overleftarrow{\mathrm{T}}\,\overleftarrow{\mathrm{T}}$.

All observables of A can then be expressed in terms of an influence functional (IF), found by performing functional integration over $F(t)$ to leave a functional of $x(t)$,

$$\mathcal{I}[x]_0^t = \mathrm{Tr}_B\left[\overleftarrow{\mathrm{T}}_B\, e^{\int_0^t \mathcal{L}(t')dt'} \rho_B(0)\right] = \left\langle e^{\int_0^t \mathcal{L}(t')dt'}\right\rangle_B. \tag{2.21}$$

The second equality is to indicate that this should be thought of as finding the expected value of $e^{\int_0^t \mathcal{L}(t')dt'}$ with respect to the distribution of quantum amplitudes $\rho_B(0)$. The IF is normally encountered when modelling open systems within the path integral formalism [8–10]. However, it should be noted that the IF as defined in Eq. (2.21) is not the same type of object as that usually called an influence functional when in the context of path integrals. The IF, $\mathcal{I}[x]_0^t$, is indeed a functional, but a functional of an operator valued function, $x(t)$, not a scalar valued function. Connection between Eq. (2.21) and conventional understanding of what constitutes an IF is made in Chap. 3.

All quantities pertaining to the behaviour of A can be calculating using the IF. For example, the reduced propagator, Eq. (2.16), can now be written

$$\Lambda_A(t, 0) = \overleftarrow{\mathrm{T}}_A\, \mathcal{I}[x]_0^t. \tag{2.22}$$

It should be clear here that $\mathcal{I}[x]_0^t$ must be considered as a functional and not a time dependent superoperator because enacting time ordering, $\overleftarrow{\mathrm{T}}_A$, requires knowledge of the how the argument function, $x(t')$, behaves at all times $t' \in [0, t]$, not just at time t. The multitime expected values of operators of A can also be expressed in terms of the IF. For example, Eq. (2.13) can be written

$$\langle X(t)X(t')\rangle = \mathrm{Tr}_A[\overleftarrow{\mathrm{T}}_A\, X(t)X(t')\,\mathcal{I}[x]_0^t \rho_A(0)], \tag{2.23}$$

provided $X(t)$ is a function of A only. As stated, this generally cannot be expressed in terms of the reduced propagator for A. The reason is that the IF is time-nonlocal and generally not divisible, $\mathcal{I}[x]_0^t \neq \mathcal{I}[x]_s^t\, \mathcal{I}[x]_0^s$. More clearly, this is the statement that the influence of B upon A across different periods of time is correlated,

$$\left\langle e^{\int_s^t \mathcal{L}(t')dt'} e^{\int_0^s \mathcal{L}(t')dt'}\right\rangle_B \neq \left\langle e^{\int_s^t \mathcal{L}(t')dt'}\right\rangle_B \left\langle e^{\int_0^s \mathcal{L}(t')dt'}\right\rangle_B. \tag{2.24}$$

Time-nonlocal correlations in the dynamics of A make it a non-Markovian system [11, 12]. In cases when time-nonlocal correlations are negligible and an effective time-local IF can be found, A is said to be Markovian. That is, A is considered Markovian if its perturbation of B induces excitations that remain in B, never to return, shown in Fig. 2.1c. The true mathematical distinction between Markovianity and non-Markovianity is not as definite as the distinction between time-locality and time-nonlocality. In fact, a precise boundary between Markovian and non-Markovian systems has not yet been established. The question of precisely what constitutes non-Markovianity is one that will not be pursued here, though a discussion on various proposed answers is found in Sect. 2.1.3. Here the term Markovian (non-Markovian) will simply be used to refer to time-locality (time-nonlocality) of the IF being used to model A.

There is now only one function, $x(t)$, passed as an argument to functionals in the effective model for A, so the explicit dependence of the IF on its argument function shall now be dropped, along with explicit dependence on integral limits, $\mathcal{I}[t] \equiv \mathcal{I}[x]_0^t$. The use of square brackets to indicate time dependence of functionals shall be used throughout, while round brackets shall be used exclusively to indicate time dependence of an operator or superoperator. Then it should be stressed that $\mathcal{I}[t] \neq \mathcal{I}(t)$ since $\mathcal{I}[t]$ and $\mathcal{I}(t)$ are fundamentally different types of mathematical object. That is, the object $\mathcal{I}(t)$ is a time-local superoperator which is found by passing the definite operator valued function $x(t)$ to functional $\mathcal{I}[t]$ as an argument and performing all time integrals. Note also that simply passing $x(t)$ to $\mathcal{I}[t]$ is generally not equivalent to performing time ordering, $\mathcal{I}(t) \neq \overleftarrow{\mathrm{T}}_A \mathcal{I}[t]$. Special cases where the equality, $\mathcal{I}(t) = \overleftarrow{\mathrm{T}}_A \mathcal{I}[t]$, does hold shall be discussed in Sect. 2.1.3.1.

2.1.1.2 Time-Nonlocal Liouvillians

The IF is a moment generating functional. That is, it is defined by its expansion in terms of superoperator valued moments of the Liouvillian [1, 13, 14],

$$\mathcal{I}[t] = \sum_{n=0}^{\infty} \frac{1}{n!} \left\langle \left(\int_0^t \mathcal{L}(t')dt' \right)^n \right\rangle_B . \tag{2.25}$$

Note that the time ordering operation is implicitly included in the definition of the angular brackets, Eq. (2.21), and has not yet been enacted. Explicitly, the nth moment in the series is given by

$$\left\langle \left(\int_0^t \mathcal{L}(t')dt' \right)^n \right\rangle_B = \int_0^t dt_n \int_0^t dt_{n-1} \ldots \int_0^t dt_1 \, \langle \mathcal{L}(t_n)\, \mathcal{L}(t_{n-1}) \ldots \mathcal{L}(t_1) \rangle_B . \tag{2.26}$$

The IF can alternatively be expressed in terms of a cumulant generating functional [14], defined as

$$\phi[t] = \ln(\mathcal{I}[t]) = \sum_{n=0}^{\infty} \frac{1}{n!} \overline{\left(\int_0^t \mathcal{L}(t')dt' \right)^n} , \tag{2.27}$$

where an overline indicates a cumulant evaluated with respect to $\rho_B(0)$. Cumulants can be calculated in terms of the moments, Eq. (2.26). For example, first moments and first cumulants are identical, while the second cumulant is related to the first and second moments via

$$\overline{\left(\int_0^t \mathcal{L}(t')dt'\right)^2} = \left\langle\left(\int_0^t \mathcal{L}(t')dt'\right)^2\right\rangle_B - \left\langle\int_0^t \mathcal{L}(t')dt'\right\rangle_B^2. \tag{2.28}$$

Both Eqs. (2.25) and (2.27) take the form of a perturbative series in the interaction potential, V. A perturbative description using the cumulant generating functional, Eq. (2.27), is nonperturbative in terms of moments, since an infinite number of moments must be summed to reproduce any exponentiated cumulant. Constructing effective perturbative models using the moment generating functional, Eq. (2.25), is possible if the interaction between A and B happens effectively instantaneously at a single point in time, such as in scattering problems [15], but here the interest is in systems for which this approximation cannot be made so the series of cumulants shall be used.

A Liouvillian can be found for A by taking the derivative of Eq. (2.27),

$$\mathcal{L}_A[t] = \frac{d\phi[t]}{dt} = \overline{\mathcal{L}(t)} + \int_0^t \overline{\mathcal{L}(t)\,\mathcal{L}(t')}dt' + \mathcal{O}(V^3), \tag{2.29}$$

such that it is possible to write

$$\mathcal{I}[t] = e^{\int_0^t \mathcal{L}_A[t']dt'}. \tag{2.30}$$

The use of square brackets, $\mathcal{L}_A[t]$, indicates that the Liouvillian is itself a functional of $x(t)$, since Eq. (2.29) involves integrals over $x(t)$ to all orders in its expansion beyond leading order. These integrals are also what makes $\mathcal{L}_A[t]$ time-nonlocal. Defining the superoperator of A,

$$x^{\pm}\rho = x\rho \pm \rho x, \tag{2.31}$$

the first cumulant of $\mathcal{L}(t)$ appearing in Eq. (2.29) is

$$\overline{\mathcal{L}(t)} = -ix^-(t)\langle F(t)\rangle, \tag{2.32}$$

where $\langle F(t)\rangle$ is the scalar valued first moment of $F(t)$, while the second is [13]

$$\overline{\mathcal{L}(t)\,\mathcal{L}(t')} = -x^-(t)\Re[C(t,t')]x^-(t') - ix^-(t)\Im[C(t,t')]x^+(t'), \tag{2.33}$$

where

$$C(t,t') = \langle F(t)F(t')\rangle - \langle F(t)\rangle\langle F(t')\rangle \tag{2.34}$$

is the scalar valued second cumulant of $F(t)$. In the case that the interaction potential is non-abelian, Eq. (2.32) should be read as a dot product between a vector of superoperators, $x^-(t)$, with a vector of scalars, $\langle F(t)\rangle$, while Eq. (2.33) should be read as a dot product between vectors of superoperators under a metric of scalars given by $C(t,t')$.

The leading order contribution to $\mathcal{L}_A[t]$ is time-local and represents a unitary driving of A, described by the mean interaction potential $\langle V(t)\rangle_B = x(t)\langle F(t)\rangle$. If only this leading order term were retained in Eq. (2.29) then $\mathcal{L}_A[t]$ could simply be written as a time-local superoperator, $\mathcal{L}_A[t] \to \mathcal{L}_A(t) = -ix^-(t)\langle F(t)\rangle$, since the time integrals that make $\mathcal{L}_A[t]$ a functional only appear beyond leading order. Note, however, that the IF in this case would still strictly be a functional since, from Eq. (2.30), it would involve an integral over the superoperator $\mathcal{L}_A(t)$. The second term in Eq. (2.29) represent a time-nonlocal self interaction of A induced by the backaction of B after previously being perturbed by A. The function, $C(t,t')$, is a measure of how B is responds at time t to perturbation at a previous time t': it is a response function. The result of B be infinite and continuous is typically that $C(t,t')$ decays to zero over some characteristic timescale $t - t' \sim \tau_c$, limiting the range of time in which system A experiences backaction in response to previously perturbing B. This condition is also generally required to justify a perturbative truncation of $\mathcal{L}_A[t]$ [11, 16]. Discarding terms $\mathcal{O}(V^3)$ is a Gaussian approximation since, by definition, the cumulants which these terms derive from vanish identically when ρ_B is a Gaussian distribution with respect to $F(t)$ [13, 17]. In this case, the retention only of terms $\mathcal{O}(V^2)$ is nonperturbative, and is exact for all coupling strengths and even if B is not infinite and continuous.

In further discussion it will be convenient to express the interaction picture Liouvillian, Eq. (2.29), in the form

$$\mathcal{L}_A[t] = -ix^-(t)\langle F(t)\rangle + x^-(t)\mathcal{F}[t], \tag{2.35}$$

with

$$\mathcal{F}[t] = -\int_0^t \Re[C(t,t')]x^-(t')dt' - i\int_0^t \Im[C(t,t')]x^+(t')dt' + \mathcal{O}(V^3). \tag{2.36}$$

Neglecting terms $\mathcal{O}(V^3)$ and assuming $\langle F(t)\rangle = 0$, the resulting IF is of Feynman-Vernon form [1, 8, 9],

$$\mathcal{I}[t] = \exp\left(-\int_0^t\int_0^{t'} x^-(t')(\Re[C(t',t'')]x^-(t'')dt' + i\Im[C(t',t'')]x^+(t''))dt''dt'\right). \tag{2.37}$$

2.1.1.3 The Harmonic Bath

The simplest possible model for a quantum environment is the harmonic bath (HB) [9, 18–20]. A HB consists of an infinite collection of harmonic oscillators whose frequencies form a continuum. The bosonic nature of harmonic oscillators make application of the HB widespread: it is appropriate for modelling both phononic and photonic environments [21, 22]. The Hamiltonian of a typical HB is

$$H_{HB} = \sum_k \omega_k b_k^\dagger b_k + \frac{1}{2}\sum_k (\alpha_k b_k + \alpha_k^* b_k^\dagger), \tag{2.38}$$

where b_k, ω_k, and α_k are the annihilation operator, frequency, and displacement of the kth mode respectively. Considering an interaction of the form Eq. (2.4) with only one term in the sum for simplicity, it is assumed that the HB couples linearly to the system upon which it will have a dissipative influence, meaning

$$F = \sum_k (g_k b_k + g_k^* b_k^\dagger), \tag{2.39}$$

where g_k determines how strongly the kth mode of the HB couples to the system. Another common assumption is that, before interaction with another system begins, the HB is in thermal equilibrium at some inverse temperature β. That is,

$$\rho_{HB}(0) = \frac{e^{-\beta H_{HB}}}{Z}, \tag{2.40}$$

where Z is the partition function. The density matrix $\rho_{HB}(0)$ is Gaussian with respect to F and so terms $\mathcal{O}(V^3)$ in the expansion of the Liouvillian, Eq. (2.29), vanish identically. The first moment of $F(t)$ and response function are then readily calculated to be

$$\langle F(t)\rangle = \Re\left[\sum_k \frac{\alpha_k^* g_k}{\omega_k}\right], \tag{2.41}$$

and

$$C(\tau) \equiv C(t, t-\tau) = \int_0^\infty J(\omega)\left(\cos(\omega\tau)\coth\left(\frac{\beta\omega}{2}\right) - i\sin(\omega\tau)\right)d\omega, \tag{2.42}$$

respectively. The quantity,

$$J(\omega) = \sum_k |g_k|^2 \delta(\omega - \omega_k), \tag{2.43}$$

is called the spectral density (SD). For a HB the relaxation timescale, τ_c, is determined by both the SD and the temperature $1/\beta$, the rule of thumb being that more structured SDs and lower temperatures lead to longer relaxation times. To see this, notice that in the zero temperature limit, $\beta \to \infty$, the autocorrelation function, Eq. (2.42), reduces to a one-sided Fourier transformation of the SD. Thus, if system B can be modelled by a HB with a flat, featureless SD then memory effects in A will be short ranged, with the Markovian limit, $C(\tau) \sim \delta(\tau)$, being reached for a SD which is constant everywhere.

The SD of a realistic system can be highly structured. When considering solid state systems for example, A could be some impurity within a crystal lattice, B,

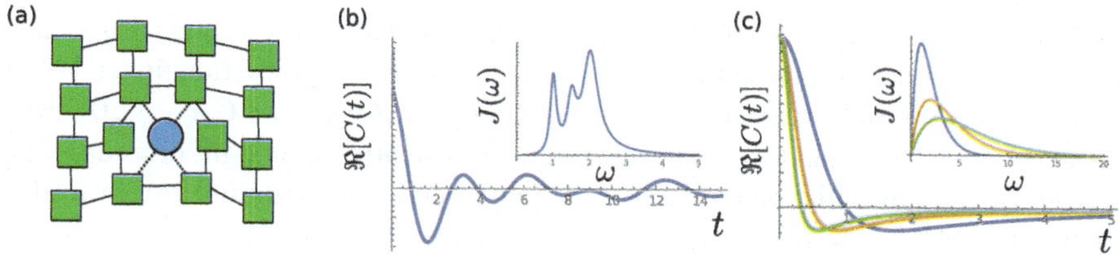

Fig. 2.2 **a** Schematic of an impurity (blue circle) interacting with a surrounding inhomogenous crystal lattice (green squares). **b** The real part of a response function (arbitrary units) resulting from a spectral density (inset, arbitrary units) representing the coupling of an impurity degree of freedom to an inhomogenous crystal lattice. In this case the impurity couples strongly to modes with three distinct frequencies, resulting in a complex form of the response. **c** The real part of a response function (arbitrary units) resulting from an Ohmic-type spectral density (inset, arbitrary units). The blue, orange and green lines correspond to three different choices of cutoff frequency in arbitrary units: $\omega_c = 1$, $\omega_c = 2$ and $\omega_c = 3$ respectively

visualised in Fig. 2.2a. Inhomogeneity of the lattice can result in a coupling of the impurity to some vibrational modes much more strongly than others [21]. The SD of B would then be structured around the frequencies of these modes, resulting in a complicated response of B to perturbation, shown in Fig. 2.2b. However, in idealised situations a simple structure is found to capture the essential dissipative influences B can have on A. If only the low energy behaviour of the environment is important, for instance, a commonly used SD is the Ohmic-type [23],

$$J(\omega) = \alpha \frac{\omega^{\nu}}{\omega_c^{\nu-1}} f_c(\omega/\omega_c), \tag{2.44}$$

where ν is the Ohmicity parameter and α is a dimensionless number which quantifies the overall coupling strength between A and B. The case $\nu = 1$ is said to be Ohmic, while $\nu < 1$ and $\nu > 1$ are sub- and super-Ohmic respectively. The dimensionless function $f_c(x)$ is a monotonically decaying cutoff function, included to ensure $J(\omega)$ dies off over a scale set by a cutoff frequency ω_c. At low energies the behaviour of A should be neutral to the specific choice of cutoff function, though for convenience a simple exponential form is often taken $f_c(x) = e^{-x}$. The SD, Eq. (2.44), therefore has just a single peak, the width of which for a given ν increases with ω_c, such that $\tau_c \sim \omega_c^{-1}$, shown in Fig. 2.2c. An idealised situation in which this form of SD has use is if B is a homogeneous crystal lattice with D spatial dimensions whose deformations A couples to such that [21] $g_k \sim \sqrt{\omega_k}$. Then B can be modelled as a HB using a SD of the form Eq. (2.44) with $\nu = D$ and a cutoff frequency $\omega_c^{-1} \sim L$ for some characteristic length scale L of A. The greater the physical extent of A, the more significantly it perturbs B, and so the longer the relaxation time τ_c. Additionally, a HB with an Ohmic SD is known to induce highly non-Markovian behaviour in A in the form of dissipative phase transitions [18] to phases in which memory effects can be infinite in range, despite a decaying response function.

Notice that in cases when B is not a HB the truncation of the Liouvillian, Eq. (2.29), to $\mathcal{O}(V^2)$ is still determined entirely by $\langle F(t)\rangle$ and a single response function, $C(t, t')$. Thus, for models of B whose response is stationary, i.e. $C(t, t') = C(t - t', 0)$, truncation of the Liouvillian to second order can equivalently be thought of as approximating B as a HB with some effective SD and effective temperature [9]. In further discussions, where A and B are talked about generically, it will therefore suffice to refer to B in terms of such effective parameters, as though B were indeed a HB.

A quantity that will be useful to refer to in future discussions is the function,

$$\eta(t) = i\lambda_c t + \int_0^t \int_0^{t'} C(t'')dt''dt', \tag{2.45}$$

where

$$\lambda_c = \int_0^\infty \frac{J(\omega)}{\omega}d\omega = \alpha\omega_c\Gamma(\nu) \tag{2.46}$$

is called the reorganisation energy and $\Gamma(\nu)$ is the gamma function. The second equality holds for a SD of the form Eq. (2.44) with an exponential cutoff function. The reorganisation energy represents the apparent shift in the overall energy of a HB that results from coupling it to another system. Given that the system and the HB are suddenly brought in contact at time $t = 0$, as implied by the initial condition Eq. (2.9), this shift in energy occurs dynamically as the oscillators of HB reorganise to find their new equilibrium configurations, different from the displaced equilibrium of the isolated HB described by the Hamiltonian H_{HB}.

2.1.1.4 The Two-Level System

The simplest possible model of a quantum system, A, is a two-level system (TLS) consisting of a ground state and an excited state with energy ω_0. The most direct physical interpretation of a TLS is as representing a single qubit of quantum information [24], making it a foundational element in the field of quantum computing. Understanding the decoherence of a TLS due to its environment is therefore of paramount importance in realising quantum computation in practice [25]. Another application of the TLS is in modelling excitonic degrees of freedom in quantum dots [26–28] which represent building block for many emerging quantum technologies [29–31]. Additionally, models for a TLS can be simulated in experiment, using trapped ions [32, 33] or superconducting circuits [34, 35] for example.

The operator σ that takes the TLS from its excited state to its ground state shall be called the transition operator of the TLS, with which it is convenient to define the excited state projection operator

$$p_\sigma = \sigma^\dagger\sigma, \tag{2.47}$$

and the transition dipole

$$x_\sigma = \frac{1}{2}(\sigma + \sigma^\dagger). \tag{2.48}$$

The expected value of the projection operator, $\langle p_\sigma(t)\rangle$, is the probability of finding the TLS in its excited state at any given time. The expected value of the transition dipole, $\langle x_\sigma(t)\rangle$, the transition dipole moment, is the real part of the transition amplitude, $\langle \sigma(t)\rangle$.

The Hamiltonian for the isolated TLS is

$$H_A = \omega_0 p_\sigma. \tag{2.49}$$

To make the TLS an open system, it shall be assumed to couple to two independent HBs, B_p and B_x (see Fig. 2.3a). The interaction potential in the lab frame, Eq. (2.8), then has the form

$$V = p_\sigma F_p + x_\sigma F_x, \tag{2.50}$$

where F_p and F_x each assume the form Eq. (2.39). Overall, this model could represent an excited state of a quantum dot that couples to its surrounding phononic environment through $p_\sigma F_p$, and whose electromagnetic dipole is being addressed by a photon field through $x_\sigma F_x$ [36]. The stationary response functions (SDs) of B_p and B_x are denoted $C_p(t)$ ($J_p(\omega)$) and $C_x(t)$ ($J_x(\omega)$) respectively, while it is taken that $\langle F_p(t)\rangle = 0$ and $\langle F_x(t)\rangle = \Omega$. In the lab frame, the exact time-nonlocal Liouvillian functional for the open TLS is then

$$\mathcal{L}_A[t] = -i\omega_0 p_\sigma^-(t) - i\Omega x_\sigma^-(t) + p_\sigma^-(t)\mathcal{F}_p[t] + x_\sigma^-(t)\mathcal{F}_x[t] \tag{2.51}$$

where $\mathcal{F}_{x/p}[t]$ is given by the leading order term in Eq. (2.36) in terms $x_\sigma(t)/p_\sigma(t)$ and $C_{x/p}(t)$. A superscript $\pm$ indicates a superoperator, as defined in Eq. (2.31). In the lab frame operators of the TLS are actually time independent, $\sigma(t) = \sigma$, but still must have associated time variable since ultimately they are to be chronologically ordered under $\overleftarrow{\mathrm{T}}$.

Various limits and special cases of Eq. (2.51) shall be used as examples in the following sections. To qualitatively illustrate the typical external influences experienced by a TLS, a single exactly solvable special case is presented here. The case considered is that the TLS is degenerate, $\omega_0 = 0$, and that the coupling to B_z vanishes, $C_z(t) = 0$. Assuming the TLS is initially in its excited state, $\langle p_\sigma(0)\rangle = 1$, the probability of finding it in this state at time t is

$$\langle p_\sigma(t)\rangle = \frac{1}{2} + \frac{1}{2}e^{-\Re[\eta_x(t)]}\cos(\Omega t), \tag{2.52}$$

where $\eta_x(t)$ is defined in terms of $C_x(t)$ and $J_x(\omega)$ as in Eq. (2.45).

Note that Ω is of order $\mathcal{O}(g_k)$ while, from Eq. (2.45), $\eta_x(t)$ is of order $\mathcal{O}(|g_k|^2)$, and that, from Eq. (2.39), Ω can be made arbitrarily large compared to $|g_k|^2$ by varying α_k independently of g_k. Then in the limit of weak coupling between the TLS

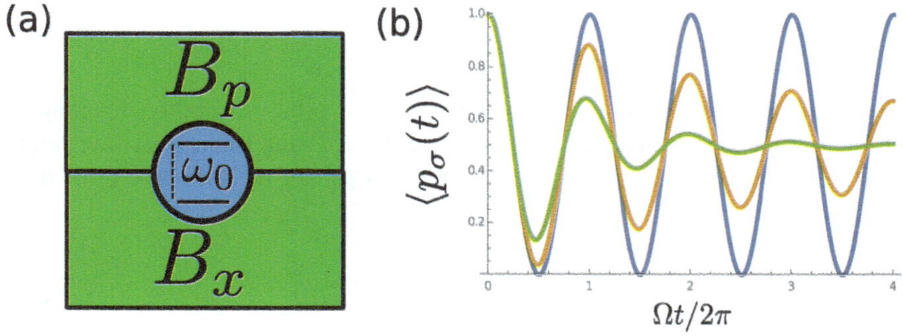

Fig. 2.3 **a** Schematic of a two-level system with energy splitting ω_0 interacting with two independent environments, B_p and B_x. **b** Rabi oscillations of the two-level system induced by B_x alone. The blue line indicates the ideal case of perfectly coherent oscillations at the bare Rabi frequency Ω, occuring in the limit $\alpha \to 0$. The orange and green lines represent the case that the Rabi oscillations are damped, for couplings $\alpha = 0.75$ and $\alpha = 3$, respectively. The cutoff frequency used here is $\omega_c = \Omega$

and B_x with large Ω, Eq. (2.52) is approximately given by

$$\langle p_\sigma(t)\rangle \approx \frac{1}{2} + \frac{1}{2}\cos(\Omega t)\,, \tag{2.53}$$

which is plotted in Fig. 2.3b. Coherent oscillations in the probability of finding the TLS in its excited state are known as Rabi oscillations [37]. The frequency of these oscillations, Ω, is known as the Rabi frequency. The induction, and understanding, of Rabi oscillations in a TLS using an external probe system is of fundamental importance, since, in theory, it provides a means by which the TLS can be initialised in a given state in experiment. Of course, the coherent oscillations described by Eq. (2.53) represent an idealised and unrealistic influence of B_x upon the TLS. For any finite coupling of the TLS to B_x there is also an irreversible, incoherent dampening of Rabi oscillations. Such damped Rabi oscillations are plotted in Fig. 2.3b for increasing coupling, where it is assumed that the SD, $J_x(\omega)$, is of the form Eq. (2.44) with $\nu = 1$ and that B_x is initially in its ground state, $\beta_x^{-1} = 0$. In this case $\eta_x(t) = \alpha \ln(1 + i\omega_c t)$ such that $\Re[\eta_x(t)]$ monotonically increases with t, diverging logarithmically, meaning that the amplitude of Rabi oscillations monotonically decrease with t and that the probabilities of finding the TLS in either of its states become equal, $\lim_{t\to\infty}\langle p_\sigma(t)\rangle = 1/2$. The coupling strength α simply scales the overall magnitude of $\Re[\eta_x(t)]$ such that increasing α results in a faster decay of Rabi oscillations.

2.1.2 *Markovian Systems*

2.1.2.1 Time-Local Liouvillians

Working in the interaction frame, the Liouvillian for a time-local d-dimensional open system can always be written in the form [38, 39]

$$\mathcal{L}_A(t) = -ih^-(t) - \sum_n \gamma_n(t)\mathcal{D}(L_n), \tag{2.54}$$

for some number of scalar decoherence rates $\gamma_n(t)$, and operators L_n. The first term generates unitary dynamics under a Hermitian operator $h(t)$

$$-ih^-(t)\rho = -i[h(t), \rho], \tag{2.55}$$

while the second term is a sum of dissipator superoperators, defined by

$$\mathcal{D}(L_n)\rho = L_n\rho L_n^\dagger - \frac{1}{2}\{L_n^\dagger L_n, \rho\}. \tag{2.56}$$

Dissipator superoperators give rise to two types of non-unitary dynamics, both falling under the guise of decoherence. If $L_n = L_n^\dagger$ then the respective dissipator simplifies to

$$\mathcal{D}(L_n) = \frac{1}{2}(-iL_n^-)^2. \tag{2.57}$$

This is the square of a generator of unitary dynamics under a Hamiltonian L_n, so it describes pure dephasing of superpositions of eigenstates of L_n. When $L_n \neq L_n^\dagger$ the respective dissipator results in incoherent population transfer in some basis as well as dephasing in that basis.

Momentarily restoring the explicit function and time dependence of functionals, the IF for time-local models is

$$\mathcal{I}[x]_0^t = e^{\int_0^t \mathcal{L}_A(t')dt'}. \tag{2.58}$$

Since this IF is defined in terms of a time-local Liouvillian it is divisible,

$$\mathcal{I}[x]_0^t = \mathcal{I}[x]_s^t\,\mathcal{I}[x]_0^s, \tag{2.59}$$

which implies divisibility of the reduced propagator,

$$\Lambda_A(t, 0) = \Lambda_A(t, s)\Lambda_A(s, 0), \tag{2.60}$$

and yields a time-local master equation for the RDM

$$\dot{\rho}_A(t) = \mathcal{L}_A(t)\rho_A(t). \tag{2.61}$$

Divisibility of the IF, Eq. (2.59), allows the expression for correlation functions, Eq. (2.23), to be rearranged in terms of the reduced propagator, validating the QRT,

$$\langle X(t_2)X(t_1)\rangle = \mathrm{Tr}_A[X(t_2)\Lambda_A(t_2,t_1)X(t_1)\Lambda_A(t_1,0)\rho_A(0)]. \tag{2.62}$$

A subtle point here is that both Eqs. (2.61) and (2.60) are implied by Eq. (2.59) but the converse is not true. The implication of this is that being able to find a time-local master equation or divisible propagator to model an open system, even if exact, does not validate the use of the QRT which only results from divisibility of the IF. A class of models, the pure dephasing models, for which the IF is not divisible but for which both a divisible reduced propagator and a time-local master equation can be obtained exactly is discussed in Sect. 2.1.3.1.

Only when $\gamma_n(t) > 0$ for all n and t is the Liouvillian given by Eq. (2.54) guaranteed to result in a physical RDM for all t and arbitrary L_n, in which case the model is also said to be Markovian. It is possible to have a physical time-local model with $\gamma_n(t) < 0$ but the additional constraints upon $\gamma_n(t)$ required to guarantee physicality are generally system specific. As an example, consider the damped Rabi oscillations of a TLS as described by Eq. (2.52). The procedure to derive the time-local Liouvillian of the form Eq. (2.54) that describes these dynamics will be made clear in following sections. Here it suffices to point out that the damping of Rabi oscillations originates from a dissipator superoperator: $-\Re[\dot{\eta}_x(t)]\mathcal{D}(x_\sigma)$. The specific choice of SD (Ohmic), used to illustrate the damping led to monotonic Markovian decay, $\Re[\dot{\eta}_x(t)] > 0$, but the general condition that must be satisfied for physicality of the TLS is actually $\Re[\eta_x(t)] > \Re[\eta_x(0)]$ for all t. This condition can be found by diagonalising the 2×2 RDM and requiring positive eigenvalues and allows for $\Re[\dot{\eta}_x(t)]$ to go negative, but only after having been positive for some time. This is known to occur at any temperature if the SD of the HB, B_x, is superohmic, of the form Eq. (2.44) with $\nu > 3$ [40]. If an environment induces such negative rates of decoherence in a quantum system it is typically considered to be having a non-Markovian influence [40]. A point of interest is that a TLS that experiences such non-Markovian damping of Rabi oscillations also lacks a notion of equilibration, since in this case $\lim_{t\to\infty}\Re[\eta_x(t)] < \infty$. This means that, after a transient damping, undamped Rabi oscillations with frequency Ω persist eternally. Then the TLS effectively evolves unitarily at large times, showing that in this case non-Markovianity is could actually be of practical use since it stabilises Rabi oscillations that might be induced deliberately in experiment.

2.1.2.2 Redfield-Secular-Markov Approximations

The Redfield approximation is to assume there is no notion of chronological order between operators at different times, such that their time ordering through $\overleftarrow{\mathrm{T}}$ is trivial.

That is, one simply ignores the functional nature of time-nonlocal Liouvillians by treating the time variable in the upper integral limit of the functional as the only time variable relevant for time ordering, making it a time dependent superoperator. To all orders in the coupling strength, and for general non-abelian interaction potentials, the Redfield approximation reads

$$\mathcal{L}_A[t] \approx \mathcal{L}_A(t). \tag{2.63}$$

The time-local Liouvillian superoperator $\mathcal{L}_A(t)$ decomposes into a unitary part and a dissipative part,

$$\mathcal{L}_A(t) = -ix^-(t)\langle F(t)\rangle + x^-(t)\mathcal{F}(t), \tag{2.64}$$

where

$$\mathcal{F}(t) = \int_0^t \Re[C(t,t')]x^-(t')dt' + i\int_0^t \Im[C(t,t')]x^+(t')dt' + \mathcal{O}(V^3). \tag{2.65}$$

Note that this definition of the superoperator $\mathcal{F}(t)$ is formally identical to the definition of the functional $\mathcal{F}[t]$ given by Eq. (2.36), but, again, it should be stressed that Eqs. (2.65) and (2.36) represent two different mathematical objects. The Redfield approximation can alternatively be expressed as approximating the IF for a given open system as time-local (and divisible),

$$\mathcal{I}[t] \approx e^{\int_0^t \mathcal{L}_A(t')dt'}, \tag{2.66}$$

such that use of the QRT can only be justified when the Redfield approximation is justified. In the case $\langle F(t)\rangle = 0$ and terms $\mathcal{O}(V^3)$ are neglected in Eq. (2.65), the time-local master equation for the RDM that results from making the Redfield approximation is the well known Redfield master equation [41].

The use of the Redfield approximation is not strictly justified by weak coupling, but rather by weak temporal correlations among the noncommuting operators that make up $x = \{x_j\}$. In other words, the Redfield approximation requires $\langle[x_{j'}(t'), x_j(t)]\rangle \approx 0$, for all j, j', t, and t'. Note that this is not satisfied in the model of a TLS given by Eq. (2.51), despite time independence of operators, since the interaction is non-abelian, $[p_\sigma(t'), x_\sigma(t)] = [p_\sigma, x_\sigma] \neq 0$. When the interaction potential is abelian with $[H_A, V] = 0$ then $\langle[x_{j'}(t'), x_j(t)]\rangle = 0$ for all j, j', t, and t' by definition, in which case the Redfield approximation becomes exact. In analogy to the concept of energy conserving potentials in classical mechanics, interactions that have the property $[H_A, V] = 0$ shall be called conservative, though the conservation is not of energy as in the classical case but rather of probabilities of finding the open system in its energy eigenstates. Models with conservative interactions, for which the Redfield approximation is exact, are known as pure dephasing models and will be discussed in detail in Sect. 2.1.3.1.

The time-local Liouvillian, $\mathcal{L}_A(t)$, obtained through the Redfield approximation, Eq. (2.63), is not necessarily in the form Eq. (2.54) required to ensure it describes a physical system. To put it into this form the non-unitary component, $x^{-}(t)\mathcal{F}(t)$, needs to be secularised. Secularisation is to assume a conservative property of the dissipation that is not quite as strong as requiring a conservative potential,

$$[H_A^{-}, x^{-}(t)\mathcal{F}(t)] \approx 0, \tag{2.67}$$

such that terms in $x^{-}(t)\mathcal{F}(t)$ that do not respect this property can be neglected. This equivalent to neglecting terms in $x^{-}(t)\mathcal{F}(t)$ that are not invariant when transformed back into the lab frame. Similar to the Redfield approximation, Eq. (2.67) holds exactly in the case of a conservative potential, but also in the case where H_A is degenerate, independent of the potential. A subtle point about secularisation is that, if nonsecular terms are discarded from $\mathcal{L}_A(t)$ and then the limit of a degenerate Hamiltonian H_A is taken, the resulting model makes incorrect predictions, since nonsecular terms become secular in this limit and their neglect is unfounded. To neglect nonsecular terms away from this degenerate limit is called the secular approximation. The overall regime of validity for the secular approximation must then be where H_A has large energy gaps, to avoid the degenerate limit, and where coupling to the environment is weak, such that interaction is only slightly nonconservative, i.e. $[H_A, V] \approx 0$.

The Redfield and secular approximations can be made to all orders of coupling strength in $\mathcal{L}_A[t]$ for arbitrary open quantum systems to obtain a time-local Liouvillian of the form Eq. (2.54). However, the time dependence of the decoherence rates, $\gamma_n(t)$, of such a Liouvillian means they are not guaranteed to describe Markovian, or even physical, systems in general. The final approximation needed to obtain a genuinely Markovian model of an open system is the Markov approximation. As stated, the appearance of negative decoherence rates is typically indicative of non-Markovian physics. The Markov approximation is to eliminate the possibility of this occurring by approximating the decoherence rates by their constant asymptotic values

$$\gamma_n(t) \approx \gamma_n = \lim_{t\to\infty} \gamma_n(t). \tag{2.68}$$

For the system to be physical requires that taking this limit produces constant, positive rates. It will be shown below that the rates obtained using the Markov approximation are indeed positive if the environment is initially in thermal equilibrium. The Markov approximation is justified if decoherence rates convergence to their asymptotic values quickly and the absolute convergence of decoherence rates to their asymptotic values is only measured relative to the smallest timescale of the problem. Thus, if the timescale for relaxation of the environment, τ_c, is the by far the smallest timescale then the asymptotic limit $t/\tau_c \to \infty$ is reached quickly and the Markov approximation is justified. The assumption of fast relaxation is independent of coupling so the Markov approximation is nonperturbative.

The set of approximations given by Eqs. (2.63), (2.67), and (2.68), the Redfield-Secular-Markov (RSM) approximations, form the foundation of the theory of Markovian open quantum systems [1]. Even though their justification is based upon an assumed weak coupling between system and environment, they can be performed to all orders of coupling strength in the expansion of $\mathcal{L}_A[t]$ to obtain a Markovian description of the system.

To make explicit how the RSM approximations are practically implemented, consider an abelian, but not energy conserving, interaction between a system and environment involving just a single Hermitian coupling operator of the system, $x_1 = x$. Then the interaction frame operator $x(t)$ can always be decomposed uniquely as

$$x(t) = \sum_{l=1}^{\frac{1}{2}d(d-1)} \left(r_l \sigma_l(t) \sigma_l^\dagger(t) + c_l \sigma_l(t) + c_l^* \sigma_l^\dagger(t) \right), \tag{2.69}$$

where

$$\sigma_l(t) = \sigma_l e^{-i\epsilon_l t}, \tag{2.70}$$

are the $\frac{1}{2}d(d-1)$ normalised transition operators between eigenstates of H_A, with associated differences in eigenvalues $\epsilon_l > 0$, r_l are real scalars, and c_l are complex scalars. For the case $d = 2$ the operator $\sigma_1 = \sigma$ is the single transition operator of a TLS, exactly as used in Sect. 2.1.1.4.

Although the operationally relevant part of the right hand side of Eq. (2.70) is time independent, it must still be associated with the time variable t appearing on the left hand side since this is the chronologically relevant piece of information required for time ordering. Momentarily using $\tau = t$ to represent the chronologically relevant time variable associated with an operator, the Redfield approximation is to let

$$\sigma_l(\tau') \approx \sigma_l(\tau) e^{-i\epsilon_l (t'-t)}, \tag{2.71}$$

where $\tau > \tau'$, for all operators appearing under integrals in the time-nonlocal functional, $\mathcal{F}[t]$. It is an approximation with respect to the chronological distinction between the operators at different times. That is, the left and right hand sides of Eq. (2.71) are operators whose associated time values relevant for time ordering, $\overleftarrow{\mathrm{T}}$, are different. The Redfield approximation, Eq. (2.71), when made upon the overall operationally relevant pieces of the $\mathcal{F}[t]$, takes the form

$$x(t)x(t') \approx x(t) \sum_{l=1}^{\frac{1}{2}d(d-1)} \left(r_l \sigma_l(t) \sigma_l^\dagger(t) + c_l \sigma_l(t) e^{-i\epsilon_l (t'-t)} + c_l^* \sigma_l^\dagger(t) e^{i\epsilon_l (t'-t)} \right), \tag{2.72}$$

where $x(t')$ has been expanded using Eq. (2.69). To stress what the approximation is here, the left hand side of this equation involves operators whose time variables, t and t', are chronologically distinct so is time-nonlocal overall, while the right hand side involves only operators with the time variable t so is time-local. The secular approximation is to then expand $x(t)$ on the right hand side of Eq. (2.72) using Eq. (2.69) and to retain only terms which involve a transition operator $\sigma_l(t)$ and its conjugate,

$$x(t)x(t') \approx \sum_{l=1}^{\frac{1}{2}d(d-1)} \left(|r_l|^2 \sigma_l(t)\sigma_l^\dagger(t) + |c_l|^2 \sigma_l^\dagger(t)\sigma_l(t) e^{-i\epsilon_l(t'-t)} + |c_l|^2 \sigma_l(t)\sigma_l^\dagger(t) e^{i\epsilon_l(t'-t)} \right). \tag{2.73}$$

Taking the environment to be a HB, the dynamical power spectrum of the HB is defined as

$$S(\omega, t) = i\lambda_c + \int_0^t C(t') e^{i\omega t'} dt', \tag{2.74}$$

which satisfies $S(0, t) = \dot{\eta}(t)$ with $\eta(t)$ as defined in Eq. (2.45). After performing the RSM approximations, the time-local Liouvillian describing the influence of the HB is

$$\begin{aligned} \mathcal{L}_A(t) &= -ih^-(t) - \lim_{t\to\infty} \Re[\dot{\eta}(t)] \sum_{l=1}^{\frac{1}{2}d(d-1)} |r_l|^2 \mathcal{D}(\sigma_l\sigma_l^\dagger) \\ &- \sum_{l=1}^{\frac{1}{2}d(d-1)} |c_l|^2 J(\epsilon_l)(N_\beta(\epsilon_l)+1)\mathcal{D}(\sigma_l) - \sum_{l=1}^{\frac{1}{2}d(d-1)} |c_l|^2 J(\epsilon_l) N_\beta(\epsilon_l)\mathcal{D}(\sigma_l^\dagger), \end{aligned} \tag{2.75}$$

with

$$h(t) = x\langle F(t)\rangle + \sum_{l=1}^{\frac{1}{2}d(d-1)} |c_l|^2 \left(\Im[S(\epsilon_l, t)]\sigma_l^\dagger\sigma_l + \Im[S(-\epsilon_l, t)]\sigma_l\sigma_l^\dagger \right) + (\Im[\dot{\eta}(t)] - \lambda_c) \sum_{l=1}^{\frac{1}{2}d(d-1)} |r_l|^2 \sigma_l\sigma_l^\dagger, \tag{2.76}$$

and where $\omega N_\beta(\omega)$ is the expected energy distribution of the HB at thermal equilibrium. Note that both $N_\beta(\omega)$ and $J(\omega)$ are positive quantities, such that the decoherence rates contained in Eq. (2.75) are all positive.

The dissipators in the first line of Eq. (2.75) cause pure dephasing in the eigenbasis of H_A at a constant rate, $\lim_{t\to\infty} \Re[\dot{\eta}(t)] > 0$. The dissipation caused by the first term in the second line is decay of higher energy eigenstates to lower energy ones, while the second term on the second line represents the reverse process. The overall dissipation caused by the second line in Eq. (2.75) thermally equilibrates the open system with respect to its bare Hamiltonian H_A at the temperature of the HB, β^{-1}. The pure dephasing dissipators do not interfere with this thermalisation, since they conserve populations of energy eigenstates, and so the overall non-unitary influence of the environment contained in $\mathcal{L}_A(t)$ is of thermal equilibration. The unitary influence of the environment, described by Eq. (2.76), is of a time dependent Lamb shift of eigenvalues of H_A along with additional unitary driving of transitions between energy eigenstates described by $\langle V(t)\rangle_B = x\langle F(t)\rangle$. The coherent driving of transitions described by $\langle V(t)\rangle_B$ competes with the non-unitary thermalising dissipation, heralding the open system a driven-dissipative system. In this case the open system does not equilibrate in the canonical thermal sense, but typically does reach a steady state.

2.1.2.3 Born-Markov Approximation

Using the property of time ordering,

$$\overleftarrow{\mathrm{T}}\, e^{\int_0^t \mathcal{L}(t')dt'} = 1 + \int_0^t \mathcal{L}(t')\, \overleftarrow{\mathrm{T}}\, e^{\int_0^{t'} \mathcal{L}(t'')dt''} dt', \tag{2.77}$$

it is possible to write the master equation, Eq. (2.7), in the form

$$\dot{\rho}(t) = \mathcal{L}(t)\rho_A(0)\rho_B + \int_0^t \mathcal{L}(t)\,\mathcal{L}(t')\rho(t')dt'. \tag{2.78}$$

Then applying $\mathrm{Tr}_B\, \overleftarrow{\mathrm{T}}_B$ one arrives at the exact, formal result

$$\dot{\rho}_A(t) = \langle \mathcal{L}(t) \rangle_B \rho_A(0) + \int_0^t \mathrm{Tr}_B[\overleftarrow{\mathrm{T}}_B\, \mathcal{L}(t)\,\mathcal{L}(t')\rho(t')]dt'. \tag{2.79}$$

The Born-Markov approximation is made upon the second term in two steps. The first step is to take $\rho(t') \approx \rho_A(t')\rho_B$, which is an assumption about negligible accumulation of correlations between system and environment. This is what typically gets called the Born approximation. The next step is to take $\rho_A(t') \approx \rho_A(t)$, resulting in the inhomogenous time-local master equation

$$\dot{\rho}_A(t) = \langle \mathcal{L}(t) \rangle_B \rho_A(0) + \int_0^t \langle \mathcal{L}(t)\,\mathcal{L}(t') \rangle_B dt' \rho_A(t). \tag{2.80}$$

The physical assumption underpinning the Born-Markov approximation is that the time for the environment to relax after being perturbed, τ_c, is much less than the time it takes for it to be significantly perturbed by the system. Then in the weak coupling limit the environment is only ever infinitesimally perturbed away from its initial state and relaxes back to it instantaneously, never becoming correlated with the system, which in turn means its backaction is immediate and there are no memory effects.

The Born-Markov approximation is perturbative, perhaps somewhat implicitly, since the resulting master equation, Eq. (2.80), involves only the first and second moments of $\mathcal{L}(t)$, not cumulants. This means that, other than in the case $\langle \mathcal{L}(t) \rangle_B = 0$, such that second moments and cumulants are equal, Eq. (2.80) cannot describe the same dynamics as a time-local model obtained by making the Redfield approximation on a time-nonlocal Liouvillian. The Redfield approximation, Eq. (2.63), amounts to nonperturbatively approximating a time-nonlocal IF as time-local, from which the ability to use the QRT, Eq. (2.14), follows. The Born-Markov approximation, on the other hand, is made directly upon the density matrix of the overall closed quantum system and always result in a perturbative equation of motion for the RDM. Remembering that a time-local master equation does not imply a time-local IF, use of the QRT cannot be justified based upon the Born-Markov approximation alone.

Despite the contrasting nature of the Redfield and Born-Markov approximations, they do result in identical models for the dynamics of a RDM when the environment under consideration is a HB with $\langle F(t)\rangle = 0$, a ubiquitous situation. In the interest of offering a critical perspective on established open systems theory, the undeniable success of theories that rely on the slightly ad-hoc Born-Markov approximation, $\rho(t') \approx \rho_A(t)\rho_B(0)$, is perhaps due to this coincidental alignment with the theoretically simpler Redfield approximation, $\mathcal{L}_A[t] \approx \mathcal{L}_A(t)$. It is also of note that the Born-Markov approximation is, in some sense, not a method of open systems theory, since it is really an approximation made on a closed system. That is, it is made upon the density matrix of a unitarily evolving closed system in order to eliminate the environment and obtain a model for an open system. This is in contrast to the Redfield approximation, which is made upon a time-nonlocal model of an open system whose environment has already been eliminated.

2.1.2.4 Markovian Two-Level System

Consider the Liouvillian for the open TLS given by Eq. (2.51). To demonstrate a simple but intractable special case of this which can be tackled using the RSM approximations, consider that the TLS decouples from B_z, meaning $C_z(t) = 0$. This is model is only solvable in the case of degeneracy, $\omega_0 = 0$, in which case the population difference dynamics are given by Eq. (2.52). Working in the interaction frame with respect to $H_A = \omega_0 p_\sigma$, the Liouvillian is

$$\mathcal{L}_A[t] = -i\Omega x_\sigma^-(t) + x_\sigma^-(t)\mathcal{F}_x[t] \tag{2.81}$$

where the interaction picture operator $x_\sigma(t)$ decomposes like Eq. (2.70) with $\sigma_1 = \sigma$, $c_1 = 1/2$, and $\epsilon_1 = \omega_0$,

$$x_\sigma(t) = \frac{1}{2}\left(\sigma e^{-i\omega_0 t} + \sigma^\dagger e^{i\omega_0 t}\right). \tag{2.82}$$

Again assuming that B_x held at zero temperature, the RSM approximations can now be made on Eq. (2.81) to get

$$\mathcal{L}_A(t) = -i\Omega x_\sigma^-(t) - i\delta(\omega_0, t)p_\sigma^-(t) + J_x(\omega_0)\mathcal{D}(\sigma(t)), \tag{2.83}$$

where $\delta(\omega_0, t)$ is the dynamical Lamb shift of the bare TLS energy. By making the Markov approximation on the time dependent Lamb shift, $\delta(\omega_0, t) \approx \lim_{t\to\infty} \delta(\omega_0, t)$, and letting $\omega_0 = -\delta(\omega_0, t)$ the lab frame Liouvillian assumes the time independent form,

$$\mathcal{L}_A = -i\Omega x_\sigma^- + J_x(\omega_0)\mathcal{D}(\sigma). \tag{2.84}$$

In this case the TLS is said to be experiencing resonant driving.

The model represented by Eq. (2.84) is the simplest possible instance of a driven-dissipative Markovian model of an open quantum system. The typical use of Eq. (2.84) is to model the spontaneous emission of a resonantly driven two-level atom into an electromagnetic vacuum, a schematic of which is found in Fig. 2.4a. In the limit of zero drive, $\Omega \to 0$, the atom simply thermalises with the electromagnetic vacuum, meaning the probability of finding it in its excited state tends to 0 asymptotically, $\langle p_\sigma(\infty)\rangle = 0$. As the transition of the atom is driven, $\Omega > 0$, Rabi oscillations in $\langle p_\sigma(t)\rangle$ occur at precisely the Rabi frequency, and exponentially decay at the rate $J_x(\omega_0)$,

$$\langle p_\sigma(t)\rangle = \langle p_\sigma(\infty)\rangle + (1 - \langle p_\sigma(\infty)\rangle)\cos(\Omega t)e^{-J_x(\omega_0)t}, \tag{2.85}$$

where $\langle p_\sigma(\infty)\rangle = \Omega^2/(|J(\omega_0)|^2 + 2\Omega^2)$. This is plotted for increasing Rabi frequency in Fig. 2.4b. The dissipator superoperator in Eq. (2.84) also causes the probability of finding the TLS in any form of superposition of its eigenstates to decay at a rate $J_x(\omega_0)/2$. Decay of populations must be accompanied by decay of coherence, since if there is zero probability of finding the atom in its excited state then there is also zero probability of finding it in superposition of its excited and ground states. Note that in case of degeneracy, $J_x(0) = 0$, unitary Rabi oscillations are recovered from Eq. (2.85) independent of coupling and the TLS does not reach a steady state, in contradiction to the exact result for this limit, Eq. (2.52). This signals a breakdown of the secular approximation, as expected.

An important theoretical prediction that is made by the model for a Markovian TLS, Eq. (2.84), relates to the incoherent transition power spectrum of the TLS. This spectrum is defined as

$$S_{RF}(\omega) = \int_{-\infty}^{\infty} \lim_{t\to\infty} \overline{\sigma(t+\tau)\sigma^\dagger(t)}e^{-i\omega\tau}d\tau, \tag{2.86}$$

where the overline indicates a scalar valued cumulant, defined in terms of the scalar valued first and second moments of the transition operator $\sigma(t)$ which can be calculated using the QRT. This results in stationary second cumulant of the form [1]

$$\lim_{t\to\infty} \overline{\sigma(t+\tau)\sigma^\dagger(t)} = \langle p_\sigma(\infty)\rangle \left(\cos(\Omega\tau)e^{-\gamma\tau} + e^{-\gamma'\tau}\right) e^{i\omega_0\tau}, \tag{2.87}$$

for some γ and γ'. The power spectrum, $S_{RF}(\omega)$, is then a symmetric triplet of Lorentzian distributions peaked at $\omega = \omega_0, \omega_0 \pm \Omega$, shown in Fig. 2.4c. This power spectrum is most commonly observed, and best understood, as a stationary fluorescence spectrum of an electromagnetically driven atom [42]. In the absence of driving, $\Omega \approx 0$, the atom only fluoresces transiently, and if it has been prepared out of equilibrium, since it approaches its ground state asymptotically, $\langle p_\sigma(\infty)\rangle = 0$. This is manifest simply as a weak incoherent broadening of its idealised monochromatic frequency response into a Lorentzian distribution. Lorentzian broadening of optical lineshapes is a generic indicator of Markovian noise in open quantum systems [6].

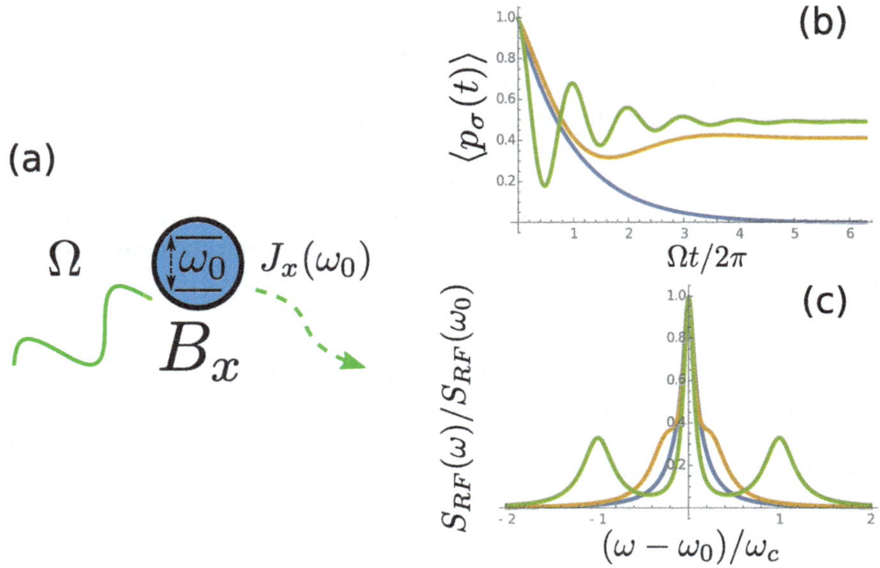

Fig. 2.4 **a** Schematic of a resonantly driven two-level atom with energy splitting ω_0 interacting with an electromagnetic vacuum, B_x. **b** A plot of the probability of finding the atom in its excited state for three different drive strengths, $\Omega = 0$ (blue), $\Omega = 0.25$ (orange) and $\Omega = 1$ (green). Parameters are chosen such that $J(\omega_0)/\omega_c = 1$. **c** A plot of the resonance fluorescence spectrum of the driven atom at three different drive strengths, $\Omega = 0$ (blue), $\Omega = 0.25$ (orange) and $\Omega = 1$ (green). Parameters used are $\gamma = 3\gamma' = 0.1\omega_c$

The splitting of the broadened monochromatic response of an energetic state of a system is a quite generic result of unitary driving of the energetic transitions of a system. On physical grounds, such driving, even when unitary, must be considered to be an effect of the system being open since it represents an explicit and persistent violation of energy conservation. In this case, it is illogical to attempt to characterise the energetic response of an open system in terms of its bare Hamiltonian H_A. Rather, it is better to describe a unitarily driven quantum system in terms of eigenstates of the effective Hamiltonian $H_A + \langle V(t) \rangle_B$, called dressed states. Although unitary driving of a quantum system is theoretically trivial it represents quite a severe external influence physically speaking, in that it is essentially an operational restructuring of the energy spectrum of the system, $H_A \rightarrow H_A + \langle V(t) \rangle_B$. Generally, each of the dressed states has overlap with all eigenstates of the bare Hamiltonian. This means measured incoherent transitional behaviour of bare energy eigenstates can have occurred rather as transitions through dressed states. In the example given above of a fluorescing atom, the emergence of two dressed states upon driving provides two additional, energetically distinct, fluorescence mechanisms by which the atom emits radiation, on top of the transitions between bare eigenstates which result in fluorescence at the bare transition energy ω_0.

2.1.3 Non-Markovian Systems

2.1.3.1 Solvable Models

To begin with, there is a class of solvable models that deserves special attention, pure dephasing models. Pure dephasing models not only allow analytic investigation of non-Markovianity, but are also experimentally realisable with exquisite control over the influence of the environment upon the system [43]. Pure dephasing models are characterised as having a conservative interaction potential, $[H_A, V] = 0$, which requires V to be abelian and means that the probabilities of finding the system in its energy eigenstates are conserved. Given that energy eigenstates of a system are extensive in number, pure dephasing models are integrable [44]. Since the probabilities of finding the system in its energy eigenstates are conserved the only influence the environment can have is dephasing of superpositions of eigenstates. The physical origin of this is the formation of entanglement between system and environment. If the system and environment are initially in an unentangled state, like Eq. (2.9), and the system is initially in a superposition, their interaction causes them to evolve into a collective superposition. This is conservative of the energy of the system, but does entangle the system with the environment. Observing the open system collapses this collective superposition, while non-observation of the environment obscures phase relationships between the possible outcomes of observations of system, manifest as dephasing.

The most general pure dephasing dynamics of the RDM of an open quantum system can be generated by a time-local Liouvillian of the form Eq. (2.54). To arrive at this conclusion, notice that all pure dephasing dynamics of the RDM can be written in terms of the d transition amplitudes of the system, $\langle \sigma_n(t) \rangle$. Arbitrary variations in the phase and magnitude of transition amplitudes can be generated dynamically using some Hamiltonian $h^-(t)$, diagonal in the energy eigenbasis, and d pure dephasing dissipator superoperators associated with the d projection operators $L_n = \sigma_n \sigma_n^\dagger$. The problem of calculating dynamics of the RDM is then reduced to a problem of finding a closed form for the time dependent eigenenergies of $h(t)$ and decoherence rates $\gamma_n(t)$. In the case of pure dephasing dynamics induced by a HB, the Liouvillian given by the first line of Eq. (2.75) can be made to describe the exact dynamics of a pure dephasing RDM by undoing the Markov approximation, i.e. no longer taking the $t \to \infty$ limit, made upon the dephasing rates.

The time-local Liouvillian describing the exact dynamics of a RDM for a given pure dephasing model is found by performing the Redfield approximation on the time-nonlocal Liouvillian for the system. For conservative interactions it is true that $[x(t), x(t')] = [x, x] = 0$ for all t and t' in both the lab and interaction frames, so there is no sense in which a functional of $x(t)$ can be put in chronological order. That is, the result of time ordering the IF of a pure dephasing model is to simply pass the time independent operator $x(t) = x$ as an argument to the IF, resulting in a time-local superoperator. The exact reduced propagator used to calculate the RDM for pure dephasing models is then simply

$$\Lambda_A(t,0) = \overleftarrow{\mathrm{T}}\,\mathcal{I}[t] = \mathcal{I}(t) = e^{\int_0^t \mathcal{L}_A(t')dt'}. \tag{2.88}$$

Note that even though the reduced propagator is divisible, the IF, $\mathcal{I}[t]$, is still time-nonlocal and thus not divisible. For pure dephasing models it is possible to calculate correlation functions of operators $X(t)$ for which $[X(t), x(t')] = 0$ using the QRT, since for these operators the time ordering in Eq. (2.23) can be trivially carried out for the same reason it can in Eq. (2.88) and divisibility of the reduced propagator would then make Eq. (2.18) an equality. However, for correlation functions of operators for which $[X(t'), x(t)] \neq 0$, time ordering cannot be trivially carried out in Eq. (2.23) and use of the QRT would only produce approximate results, even though the reduced propagator is exactly divisible.

One proposed characterisation of non-Markovianity is by observed deviation from predictions made using the QRT [45]. From the above discussion, if non-Markovianity were measured by strict analytic deviation from QRT predictions then all open systems would be non-Markovian, since even for systems with an exactly divisible reduced propagator there are correlation functions that cannot be calculated using the QRT [46]. A more practical and interesting way to diagnose non-Markovianity in pure dephasing models is to limit the scope of analyses to the evolution of the state of system; the tendency of the open system to develop into superpositions of energy eigenstates rather than to decohere, occurring when decoherence rates become negative for some time, can be understood as a non-Markovian phenomenon. There are complimentary physical prescriptions for understanding the nature of such recoherence of an open quantum system. In an information theoretic sense, perturbation of the environment by the system and the resulting backaction can be thought of as a flow of information [47, 48]. Information about an initial superposition state of the open system is regained if it develops back towards this state during its evolution, meaning non-Markovianity can be systematically witnessed using quantitative measures of the information content of an open system. Recurrence of superposition states can also be understood as a loss of entanglement between system and environment. Modelling a measuring apparatus as an ancilla system, loss of system-environment entanglement can occur if entanglement develops between the system and the ancilla [49]. This suggests observability and control of non-Markovianity through the ancilla. The physical nature of non-Markovianity in pure dephasing models can be ascribed to both information flow and entanglement, both of which admit an analysis using tools of resource theory [50–52]. Indeed, it has been shown that non-Markovianity can provide a resource with which to improve information processing [53, 54].

Pure dephasing models represent a rare class of model that are solvable directly once a time-nonlocal Liouvillian is known. There are some models where the behaviour of the open system is instead deduced by solving the model of the collectively closed system and environment first. One instance of such a model is the many-mode Jaynes-Cummings model [55, 56]. This is a model of a TLS undergoing non-abelian interaction with a HB, $V = \sum_k(\sigma g_k b_k + \sigma^\dagger g_k^* b_k^\dagger)$. By calculating the exact evolution of both the TLS and the HB and then eliminating the HB, both a time-nonlocal master equation and a time-local Liouvillian of the form Eq. (2.54) can be found to model the RDM of the TLS [56]. An interesting feature of non-abelian interactions can be isolated in this model. In the limit that the HB consists of a single oscillator whose frequency matches the energy of the TLS, the dynamics of the RDM are described by a single dissipator superoperator, $-2|\tan(|g_1|t)|\mathcal{D}(\sigma)$. This is a slightly pathological way to express the dynamics of the RDM since the rate diverges periodically, but the same form of dissipation is found with a finite decoherence rate when generalised to the many-mode case. The limit of the HB consisting of a single mode is highly non-Markovian when taken at face value, since it describes eternal undamped but incoherent oscillations of the TLS due to perfect coherent exchange with the single mode of the HB. The decoherence rate, $-2|\tan(|g_1|t)|$, is nonperturbative in the coupling, rather than being $\mathcal{O}(V^2)$ as the exact time-nonlocal Liouvillian for this model is. It is interesting that the representation of nontrivial non-Markovian behaviour in terms of a time-local Liouvillian requires a description which is nonperturbative in coupling, even though the exact influence of the HB is nominally $\mathcal{O}(V^2)$. This highlights that, simply by virtue of being time-nonlocal, Liouvillians of open systems in contact with a HB encode a highly nontrivial physical influence on the open system, despite having a simple closed form.

The final class of solvable models that will be mentioned here are those in which both system and environment are composed of harmonic oscillators, solvable using the technique of Laplace transforms, for example [57]. When the system consists of a single oscillator, the model is known to result in time-nonlocal Brownian motion of the oscillator [58–60]. Though an approximate time-local description of this Brownian motion can be found [18], it has been shown that the model must be time-nonlocal to some degree if it is to be physical [60]. Allowing for more than one oscillator to be present in the open system, an unphysical time-local model (that can result in nonpositive RDMs) can be constructed using the Redfield approximation that outperforms the physical model obtained by further performing the secular approximation [57]. If a physical time-local model that makes accurate predictions about the system cannot be found, then one must conclude that it is necessary to use a time-nonlocal, non-Markovian model. Thus, in this case the success of an unphysical time-local model for a system over physical ones implies that the system is non-Markovian.

2.1.3.2 Change of Frame

The first strategy in going beyond integrable models is to make a unitary transformation on the model of the closed system and environment, revealing a new frame to work in where the corresponding open system model becomes amenable to the RSM approximations. Unlike the unitary transformation used to move into the interaction frame, Eq. (2.6), these kind of transformations generally result in a loss of physical information about the open system in question. That is, the transformed model for the open system does not contain enough information to calculate the same set of observables as for the untransformed model. Consider a pure dephasing model with a HB environment. A polaron transformation [21, 61] can be made to find a frame in which the system and environment completely decouple, making the open system identically time-local and unitary. The frame found by making this transformation is called the polaron frame, so-called because the polaron transformed energy eigenstates of the system are hybridisations of systemic and environmental degrees of freedom, referred to as polaron dressed states. Note that the terminology of dressed states here refers to eigenstates of the closed system and environment, rather than effective energy eigenstates of a driven-dissipative open system.

Suppose that in the above example the observer introduces a weak driving of transitions between lab frame eigenstates of the open system, breaking exact integrability the model, $[H_A, V] \approx 0$. In this case, transforming into the polaron frame succeeds in decoupling the HB from eigenstates of the polaron transformed system, but at the expense of introducing nonlinear, non-abelian coupling to energetic transitions. Upon identification of an appropriate interaction frame in which to isolate this nonlinear coupling, the HB can be eliminated to obtain a time-nonlocal Liouvillian describing the dynamics of the polaron transformed system. Nonlinearity of the interaction between the system and the HB in the polaron frame means that the HB is no longer Gaussian with respect to this interaction, such that truncation of the Liouvillian to second order is no longer exact. However, the magnitude of the nonlinear polaron frame coupling is proportional to the strength of the driving induced by the observer which is small by construction. Then truncation of the Liouvillian is perturbative with respect this driving strength but nonperturbative with respect to the lab frame coupling. Upon truncating the polaron frame Liouvillian to second order, the structure of the resultant effective SD is generally broader than that of the HB in the lab frame, meaning that relaxation of the HB in the polaron frame is faster than in the lab frame, making the effects of time-nonlocality less prominent in the polaron frame. The objective this technique achieves is the identification of a frame in which the RSM approximations can be justified when it is perhaps not so in the lab frame [62].

That an effective time-local model for an open system can be found by making a polaron transformation does not imply that it is a Markovian system, since such a Markovian model provides incomplete information about the system. Indeed, observables of the system can be calculated in terms of observables of both the polaron transformed system and polaron transformed environment [36], the results of which do exhibit non-Markovian behaviour [63, 64]. One could argue that further making these mean field approximations is a strategy for solving non-Markovian models to be discussed in the next section, since it explicitly requires knowledge of the degrees of freedom of the environment as well as the system to which it couples. The combination of working in the polaron frame and performing mean field approximations can also be used to arrive at entirely nonperturbative results [65].

By performing a polaron transformation upon a closed model of a system and its environment, time-local models can be used to model non-Markovian influences of the environment. However, working in the polaron frame as described above can result in pathologies. In particular, if a system is coupled weakly to a HB in the lab frame, the nonlinear interaction potential in the polaron frame can be made linear through a perturbative expansion in this coupling. Upon eliminating the environmental degrees of freedom this results in an effective SD $J_P(\omega) \approx J(\omega)/\omega^2$, a result that is derived in Sect. 4.3. This makes truncation of $\mathcal{L}_A[t]$ in the polaron frame questionable for a SD of the form Eq. (2.44) with $\nu < 2$. Both in managing this problem and in surpassing the limit of a weakly driven system (required to justify truncation of the Liouvillian in the polaron frame), one can identify a continuum of frames [66–68], with the polaron and lab frames being limiting cases. An optimal frame in which to work can then be identified through known properties of the lab frame system and minimisation principles. This kind of technique succeeds the concept of the polaron frame but effectiveness is once again limited, particularly by the form of the lab frame SD.

2.1.3.3 Boundary Extension

When exactly solving a time-nonlocal model is not possible and no frame can be found that admits solution of an effective time-local model, one must seek strategies to tackle time-nonlocality more directly. The physical nature of how an open system perturbs its environment and the resulting backaction is complex. However, given that the environment is infinite in extent then, notionally, perturbations which return to the open system have remained close to the boundary between system and environment. This suggests that it may be possible to move this boundary, through a redefinition of which degrees of freedom are systemic and environmental, such that perturbations in the environment, and thus non-Markovianity in open systems, can be modelled explicitly. The obvious logical conclusion of such an approach is the extension of the system to encompass all of the environment, in which case one is modelling a closed system that is infinite in extent, and Markovian by definition. Both pure dephasing models with a HB and the many-mode Jaynes-Cummings model are solvable this way [1, 56].

The motivation and effectiveness of boundary extension as an approach to tackling general time-nonlocal models is best shown through example. Consider that the SD of a HB environment is approximately a Lorentzian distribution. This could occur in the situation of an impurity in an inhomogenous crystal lattice whose inhomogeneity results in strong coupling to a single specific vibrational mode. As discussed and shown in Sect. 2.1.1.4, Lorentzian broadening of the natural frequency response of an open system is an indicator of Markovian noise. This suggests the HB could be approximated as a single harmonic oscillator, whose broadened frequency response is due to the Markovian influence of another environment. Indeed, a HB whose SD is exactly a Lorentzian distribution can be exactly represented as just a single oscillator undergoing Markovian decay [69]. However, such a situation is ultimately unphysical since a physical SD must satisfy $J(\omega \leq 0) = 0$ meaning a realistic SD cannot be exactly Lorentzian. Approximation of a HB by a single mode coupled to a residual environment can be formally achieved by identifying a new set of normal modes to use as basis modes of the HB, only one of which, the reaction coordinate [70, 71], is represented by the operator F in the potential $V = xF$. The bare Hamiltonian of the HB then assumes the form of a coupling between F and the rest of the new normal modes as well as intermode coupling. The reaction coordinate can then be taken in as an additional degree of freedom of the open system. The residual HB that the reaction coordinate couples to can be then be eliminated and treated using the RSM approximations [72–74], provided that it does indeed have a Markovian influence.

The effectiveness of the procedure outlined above is constrained by the SD of the residual HB, which in turn is determined by the bare SD of the full HB. When the bare SD has a particularly simple structure the effective SD of the residual bath can be inferred exactly. For instance, a HB whose SD is approximately Lorentzian can be shown to map to a reaction coordinate coupled to a residual HB whose SD is Ohmic [72], which is known, and was shown in Sect. 2.1.1.4, to result in Markovian noise. Sometimes the SD for the residual HB cannot be inferred exactly, so the reaction coordinate decomposition itself becomes an approximation. An alternative approach is to find a new frame to work in which the HB exactly assumes the form of a chain of interacting harmonic oscillators. The benefit of this approach is that the explicit mapping between properties of the HB and properties of the chain can be found exactly [75, 76]. This chain of harmonic oscillators can be truncated, with efficient numerical techniques enabling the elimination of resultant error [77], providing a means to model non-Markovian dynamics of the system exactly [78, 79].

2.1.3.4 Augmentation

The final strategy to tackle time-nonlocality discussed in this section is to augment the open system in question with auxiliary degrees of freedom, within which time-nonlocal effects can be embedded. Methods that follow this approach are rather diverse and often numerical. Indeed, essentially all numerical techniques operate on the principle that a difficult problem can be made tractable by embedding the difficulty into degrees of freedom that are manageable by machine.

A generic physical motivation for augmentation is difficult, though the basic philosophy is that time-nonlocal systems can be made time-local through embedding into a larger system, which is intuitive since time-nonlocality in an open system is only apparent because of its embedding in an environment. A particularly intuitive, but quite specific, way to realise this approach is to treat instances of an open system at different moments in time as genuinely distinct systems. These systems can be gathered into a single augmented system, the state of which is then representable by an object similar to a density matrix of many-body system [80, 81]. Time-nonlocality can then be modelled as interactions between subsystems of this augmented system, which in turn can be accounted for in a time-local way.

The method of augmentation described above is usually realised within the path integral formalism, through time discretisation of functional integrals of scalar influence functionals, and is powerful enough to allow numerical calculation of arbitrary observables of open systems [82, 83]. The shortfall is that the size of the problem grows exponentially with the total number of discretisation timesteps, although approximations can be made which allow for scaling of the problem only with the relaxation time of the environment [80, 81]. There are many variations on this method of modelling time-nonlocality using path integrals [84–93] with a wide range of applications. For instance, path integral methods have been used to investigate nontrivial features in the emission spectra of quantum dots [94] as well as in the behaviour of cavity photons that can couple to a quantum dot [89]. Such methods have also proven effective in modelling open systems with continuous degrees of freedom [95], and those whose environment induces long-ranged time-nonlocality [96]. The exact nature of path integral methods means they have also found use in benchmarking analytic methods [66, 97].

If the IF is known and the RDM of a system can be calculated exactly over the initial period of time it takes the environment to relax, this information can be used to calculate further dynamics of the system [98–100]. Then by calculating the RDM at a finite number of time points they can be collected into a finite augmented system which can be used to for further calculations. This method of augmentation potentially enables highly efficient investigation of non-Markovian systems, but relies on some other efficient method being available to calculate the initial set of RDMs that make up the augmented system.

Augmentation can also be realised by defining a hierarchy objects similar to RDMs that are coupled together and obey a hierarchy of equations of motion [101–103]. Such an approach can be derived directly from the IF of an open system [104] and is powerful enough to address the non-Markovian phenomenon of localisation [105], but is still limited by the form of SD of the environment.

2.1.3.5 Pure Dephasing Two-Level System

Consider the TLS Liouvillian, Eq. (2.51), in the case that the TLS entirely decouples from B_x, such that $\Omega = C_x(t,t') = 0$. This is the independent boson model [21], which could be used to model the pure dephasing of an two-level impurity whose excited state couples to a surrounding crystal lattice as described in Sect. 2.1.1.3. The SD of B_p is $J_p(\omega)$ and in the case that the impurity couples to lattice deformations this has the form Eq. (2.44), where the Ohmicity parameter, ν, represents the spatial dimension of the lattice. This is depicted in Fig. 2.5a. The lab frame time-nonlocal Liouvillian for the model is

$$\mathcal{L}_A[t] = -i\omega_0 p_\sigma^-(t) - p_\sigma^-(t)\mathcal{F}_p[t]. \tag{2.89}$$

In the interest of isolating the dynamical influence of the HB on the TLS it is assumed that $\omega_0 = -\lambda_c$, where λ_c is the reorganisation energy of B_p. Then the time-local Liouvillian that determines the exact decoherence dynamics of the RDM in the lab frame is

$$\mathcal{L}_A(t) = -i\lambda_p(t)p_\sigma^- - \gamma_p(t)\mathcal{D}(p_\sigma) \tag{2.90}$$

with the decoherence rate $\gamma_p(t) = \Re[\dot{\eta}_p(t)]$ and dynamical Lamb shift $\lambda_p(t) = \Im[\dot{\eta}_p(t)]$.

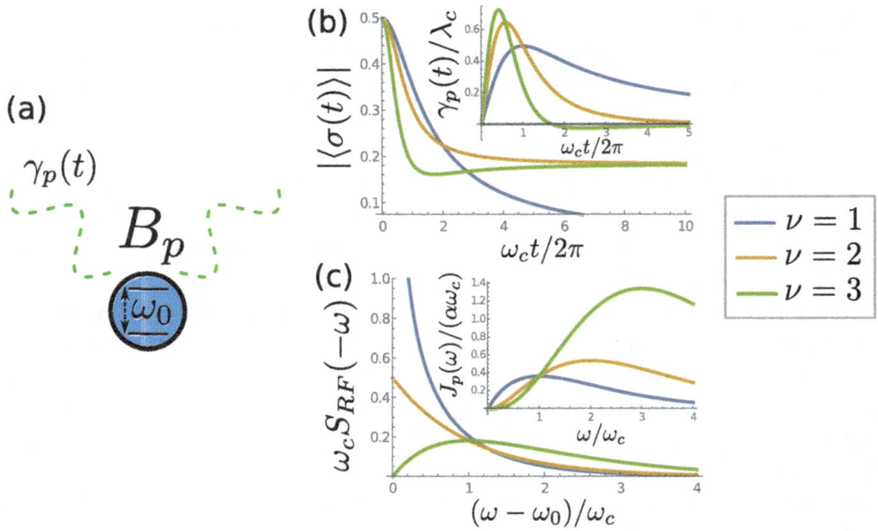

Fig. 2.5 **a** Schematic of a two-level impurity with energy splitting ω_0 undergoing pure dephasing due interaction with a surrounding crystal lattice, B_p. **b** A plot of the evolution of the magnitude of the transition amplitude for lattices of varying spatial dimension: $\nu = 1$ (blue), $\nu = 2$ (orange) and $\nu = 3$ (green). The inset shows a plot of the respective decoherence rates. Parameters used are $\alpha = 1$ and $\beta_p^{-1} = 0$. **c** A plot of the resonance fluorescence spectrum (arbitrary units) of the impurity for lattices of varying spatial dimension: $\nu = 1$ (blue), $\nu = 2$ (orange) and $\nu = 3$ (green). The inset shows a plot of the respective spectral densities of the lattice. Parameters used are $\alpha = 0.5$ and $\beta_p^{-1} = 0$

Assuming the impurity is initialised in a pure and symmetric superposition of its two states, $\langle\sigma(0)\rangle = 1/2$ and $\langle p_\sigma(0)\rangle = 1/2$, the decoherence dynamics are quantified through the magnitude of the transition amplitude,

$$|\langle\sigma(t)\rangle| = \frac{1}{2}e^{-\int_0^t \gamma_p(t')dt'}. \tag{2.91}$$

The decoherence dynamics described by Eq. (2.91) are plotted in Fig. 2.5b with varying lattice dimension, ν. The zero temperature limit, $\beta_p^{-1} \to 0$, has been taken to clarify the emergence of non-Markovianity. The appearance of negative values of $\gamma_p(t)$ with increasing ν is shown in the inset of Fig. 2.5b, and the line marking $\gamma_p(t) = 0$ has been boldened to highlight how $\gamma_p(t)$ cuts this line for $\nu = 3$.

The nature of the oscillatory component of $\gamma_p(t)$, which determines how and when it goes negative, is determined by the phase argument of $\dot{\eta}(t)$, given by $\arg(\dot{\eta}_p(t)) = \nu \arctan(\omega_c t)$. Oscillations in $\gamma_p(t)$ are sinusoidal with this phase. The arctangent is bounded from above by $\pi/2$ as $t \to \infty$, so $\gamma_p(t)$ can only go negative when $\nu > 2$, allowing the completion of at least a half period of sinusoidal oscillation in finite time. Not only does $\gamma_p(t)$ go negative when $\nu > 2$, for $k < \nu < k+1$ with even k it approaches its asymptotic value, $\lim_{t\to\infty} \gamma_p(t) = 0$, from below. The monotonic damping of oscillations of $\gamma_p(t)$ is given by the magnitude $|\dot{\eta}_p(t)| = \lambda_c(1+\omega_c^2 t^2)^{-\nu/2}$. Notice that the interaction strength, quantified by the dimensionless coupling constant α, only affects the magnitude of the decoherence rate through λ_c, so is unrelated to non-Markovian recoherence of the the impurity. Concurrent with the transition to a non-Markovian regime for $\nu > 2$ is the stabilisation of coherence of the impurity, meaning that $|\langle\sigma(t)\rangle|$ remains finite asymptotically and that there is finite probability of finding the impurity in a coherent superposition for all time. In this regime, the steady state value of $|\langle\sigma(t)\rangle|$ is determined by $\lim_{t\to\infty}\Re[\eta(t)] = \Gamma(\nu-1)/\Gamma(\nu)$, whereas for $\nu = 1$, $\lim_{t\to\infty}\Re[\eta(t)]$ diverges and the impurity completely decoheres.

To understand the nature of non-Markovian recoherence in the independent boson model, and for reference in future discussion, it is useful to introduce the trace distance [47] between two operators, X_1 and X_2, defined as

$$D(X_1, X_2) = \frac{1}{2}\mathrm{Tr}\sqrt{(X_1 - X_2)(X_1 - X_2)^\dagger}. \tag{2.92}$$

The trace distance is a scalar measure of the absolute difference between two given operators and can be used as a measure of distinguishability of two given instances of an RDM of an open system, $\rho_A(t)$ and $\rho'_A(t)$. If these RDMs represent pure and orthogonal states of a system then $D(\rho_A(t), \rho'_A(t)) = 1$, meaning they are perfectly distinguishable states. If the trace distance between two RDMs decreases as they evolve, $D(\rho_A(t), \rho'_A(t)) < D(\rho_A(0), \rho'_A(0))$, they are becoming less distinguishable and information about their initial configuration is lost. An increasing trace distance means an increase in distinguishability between the distinctly prepared quantum states, which can be interpreted as a backflow of information about the initial preparation. A good characterisation of non-Markovianity is by the pair of initial RDMs that maximises the total information backflow [47, 48]. In the independent boson model any pair of pure and orthogonal superposition states of the TLS achieve this maximisation, resulting in $D(\rho_A(t), \rho'_A(t)) = 2|\langle\sigma(t)\rangle|$. Thus, the dynamics of the TLS as described by Eq. (2.91) and plotted in Fig. 2.5b can be interpreted as loss of recoverable information about the initial state, into the HB, which later returns to the TLS if $\nu > 2$.

The transition amplitude, Eq. (2.91), can also be related to the entanglement of the TLS with the HB. For instance, the trace distance between a state of the TLS that was initially a pure superposition and the maximally mixed state is $|\langle\sigma(t)\rangle|$. Given the simplicity of the independent boson model, and the assumption that both the TLS and the HB are initially in pure states, the only way the TLS can reach its maximally mixed state is by becoming maximally entangled with the HB. The fact that non-Markovian recoherence of the TLS is accompanied by stability of superposition states therefore implies a bound on how entangled it becomes with the HB, while the approach of $\gamma_p(t)$ to zero from below indicates that this bounded level of entanglement is approached from above. In the situation that the TLS is an impurity interacting with a crystal lattice this could be interpreted as the formation of a localised region of entanglement in the vicinity of the impurity. That this only occurs for a lattice in two spatial dimensions and higher can be reasoned by noting that excitations induced by the impurity in a lattice with $\nu = 1$ (a one dimensional chain of sites) retain their intensity as they propagate through the lattice. This allows lossless propagation of information about the state of impurity throughout the lattice, manifest as long ranged entanglement which becomes maximal in the limit $t \to \infty$.

Another way to characterise the behaviour of the TLS predicted by the independent boson model is through the incoherent transition power spectrum, Eq. (2.86). In the example of a driven-dissipative TLS system given in Sect. 2.1.2.4 the spectrum, $S_{RF}(\omega)$, was interpretable as the fluorescence spectrum of an electromagnetically driven two-level atom. In the present example it is hard to justify an interpretation of $S_{RF}(\omega)$ as a stationary fluorescence spectrum of the impurity, since the impurity is energetically stable. To allow for the interpretation of $S_{RF}(\omega)$ as an emission spectrum it is now assumed that the impurity couples to an electromagnetic vacuum into which it can spontaneously emit radiation. The decay rate is assumed infinitesimally small such that spontaneous decay of the impurity is expected to occur long after any transient evolution of the overall system, meaning that no formal changes need be made to the Liouvillian of the model.

The time-nonlocal nature of the IF of the independent boson model means the QRT cannot be used to calculate the cumulant, $\overline{\sigma(t+\tau)\sigma^\dagger(t)}$, though it can readily be found by other means [106, 107]. Retaining the assumption that the HB is initially at zero temperature, but now allowing for an arbitrary initial state of the impurity, the result is

$$\lim_{t\to\infty} \overline{\sigma(t+\tau)\sigma^\dagger(t)} = \langle p_\sigma(0)\rangle e^{-\phi(0)} \left(e^{\phi(\tau)} - 1\right), \tag{2.93}$$

where

$$\phi(t) = \int_0^\infty \frac{J_p(\omega)}{\omega^2} e^{-i\omega t} d\omega. \tag{2.94}$$

Upon expanding $e^{\phi(\tau)}$ as a power series in $\phi(\tau)$, the transition power spectrum is

$$S_{RF}(\omega) = \langle p_\sigma(0)\rangle e^{-\phi(0)} \left(\frac{J_p(|\omega|)}{\omega^2} + \mathcal{O}(V^4)\right) \theta(\omega_0 - \omega), \tag{2.95}$$

where $\theta(x)$ is the step distribution. This spectrum is plotted in Fig. 2.5c for varying lattice dimension ν in the weak coupling limit such that terms $\mathcal{O}(V^4)$ contribute negligibly. The inset in Fig. 2.5c is a plot of the corresponding SDs for lattices of varying spatial dimension.

Since the impurity is not being energetically driven and the lattice is held at zero temperature, emission is only at energies less than its coherent response frequency, ω_0, signalled by the step distribution in Eq. (2.95). The collective state of the impurity and lattice at $t = 0$ is not an energy eigenstate of the overall system and generally has overlap with all such eigenstates. Thus, although the probability of finding the impurity in a given eigenstate of H_A is conserved (up to an infinitesimally slow decay) the probability of finding the impurity and lattice collectively in a given eigenstate of H is not. The eventual transition of the impurity to its ground state due to its coupling to the electromagnetic vacuum, and the resulting emission of radiation, is then influenced by collective energetic transitions between eigenstates of H. This transitional behaviour takes the form of an energetic reorganisation of the lattice in response to suddenly being brought in contact with the impurity at $t = 0$. In turn, this manifests as the development of potential energy, $\lim_{t\to\infty} \mathrm{Tr}[V\rho(t)] = \langle p_\sigma(0)\rangle\lambda_c$ where $\rho(t)$ is the collective state of impurity and lattice. When the impurity decays and emits radiation this potential energy is released, and indeed the total expected energy of emitted radiation, found by integrating $\omega S_{RF}(\omega)$ over all ω, is equal to $\langle p_\sigma(0)\rangle\lambda_c$ up to a constant normalisation.

The shape of $S_{RF}(\omega)$ is determined by the magnitude of the overlaps between the collective initial state at $t = 0$ and the eigenstates of H. That is, it is determined by which eigenstates of H the collective initial state of impurity and lattice is most likely to transition to as the lattice is reorganising. Eigenstates of H can be labelled according to the wavevectors, k, of the harmonic oscillators in the HB, such that the overlap of the collective initial state with a given eigenstate of H goes like $g_k/\omega_k + \mathcal{O}(g_k^2)$. Homogeneity of the lattice means $\omega_k \propto c|k|$, where c is the speed of sound in the lattice, such that, in the absence of additional polarisation degrees of freedom, each eigenstate of H is ν-fold degenerate for a lattice with spatial dimension ν. The overall contribution to the emission of the impurity at a single given energy, ω, is then found by taking the square magnitude and summing over all the overlaps with eigenstates of H corresponding to the energy eigenvalue ω. To leading order in g_k the result is $\sum_{|k|=\omega/c} |g_k/\omega_k|^2 = J_p(\omega)/\omega^2$, which is the leading order contribution in Eq. (2.95) as expected. It was found from the decoherence dynamics of the impurity, Eq. (2.91), that non-Markovian behaviour occurs only for $\nu > 2$. The incoherent emission of the impurity as just described is intrinsically non-Markovian independent of ν however, since it directly results from dynamical evolution of the lattice and not the impurity.

2.1.4 Summary

A review of the foundational concepts, theoretical framework, and paradigmatic models used in the field of open quantum systems has been given. The general approach one takes to mathematically model a quantum system interacting with an environment and the typical features expected of such an open system were discussed in Sect. 2.1.1. It was shown in Sect. 2.1.1.1 how an effective model for an open system is found by eliminating the degrees of freedom of the environment from a model of the collectively closed system and environment. This results in an description of the open system in terms of a time-nonlocal Liouvillian, whose derivation was given in Sect. 2.1.1.2. The prototypical model for a quantum environment, the harmonic bath, was then introduced in Sects. 2.1.1.3 and 2.1.1.4 it was demonstrated what influences such an environment can have on a two-level system.

The theory of Markovian open quantum systems, those which are time-local, was presented in Sect. 2.1.2. Properties of the models for such systems were given in

Sect. 2.1.2.1 and the set of approximations that can be used to obtain time-local models from time-nonlocal ones were given in Sects. 2.1.2.2 and 2.1.2.3. This theory was then applied to a two-level system interacting with a harmonic bath in Sect. 2.1.2.4. Those systems for which a time-local description does not suffice are non-Markovian systems, which were discussed in Sect. 2.1.3. It is not known generally how to solve time-nonlocal models but some are tractable, examples of which were given in Sect. 2.1.3.1. Beyond exact analytic solutions, a variety of different methods have been developed to tackle non-Markovian systems. The use of unitary transformations to make problems more amenable to analytic investigation was discussed in Sect. 2.1.3.2 and the idea that part of the environment could be included in the description of the system was shown in Sect. 2.1.3.3. The discussion in Sect. 2.1.3.4 was on the concept that an open system can be augmented with extra degrees of freedom in order to model non-Markovian effects. In Sect. 2.1.3.5 a time-nonlocal model for a two-level system was solved exactly to illustrate typical non-Markovian behaviour of an open system.

2.2 Tensor Networks

Diagrammatic representation of mathematical objects can provide insight both into the structure of said objects, and into how they are manipulated when performing calculations. A tensor network (TN) is a graph whose vertices represent mathematical objects called tensors and whose lines represent binary operations between tensors called contractions [108]. Here a very basic introduction to the parts of TN theory required for future discussions is given. In Sect. 2.2.1 the concepts of a tensor and the pictorial notation used to build up TN diagrams is introduced. A common class of TN that has found particularly widespread use, the matrix product [108, 109], is then discussed in Sect. 2.2.2. Calculation of a tensor by means of a TN diagram involves contracting together the tensors in the diagram, often numerically: the established way of efficiently contracting matrix product TN diagrams is presented in Sect. 2.2.3.

2.2.1 Notation

The definition of a tensor varies depending on the context; here a tensor is taken to be a multidimensional function whose arguments are scalars, and which returns either a scalar or an operator. It will be assumed here that the arguments of all tensors under discussion take values from a discrete and finite set of scalars. The mathematical object that a tensor returns for a given set of arguments need not be specified currently.

A TN is typically composed of many tensors, so it is convenient to use a shorthand notation for the arguments of such tensors. For example, a tensor with four arguments, $A(i, j, k, l)$, shall be written as

$$[A]_{ijkl} \equiv A(i, j, k, l). \tag{2.96}$$

A tensor contraction generally refers to summing over a given argument. A contraction between tensors involves multiplying them together and summing over a common argument, producing another tensor. For example one possible contraction between $[A]_{ijkl}$ and another tensor, $[B]_{mn}$, would be

$$[C]_{ijkn} = \sum_l [A]_{ijkl}[B]_{ln}, \tag{2.97}$$

where the sum is over all possible values that l can take. Clearly the ability to contract $[A]_{ijkl}$ and $[B]_{mn}$ in this way requires the arguments l and m to take on values from the same set of scalars. If the arguments of the tensors were continuous instead of discrete the sum would be replaced by an integral.

Diagrammatically, a single tensor is represented by a single vertex with as many protruding lines as the tensor has arguments. The tensor $[A]_{ijkl}$ can then be written as

$[A]_{ijkl} =$ 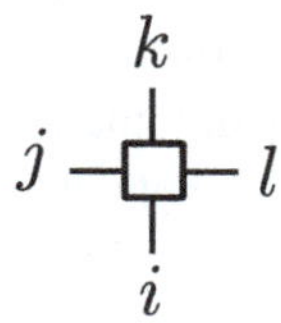

(2.98)

and similarly

$[B]_{mn} =$ 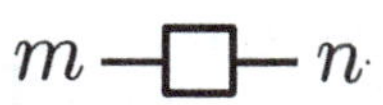

(2.99)

The protruding lines will be referred to as legs, such that one would say A is a 4-leg tensor. The contraction, Eq. (2.97), is represented as a TN by connecting together legs of $[A]_{ijkl}$ and $[B]_{mn}$ like so

$[C]_{ijkn} =$ 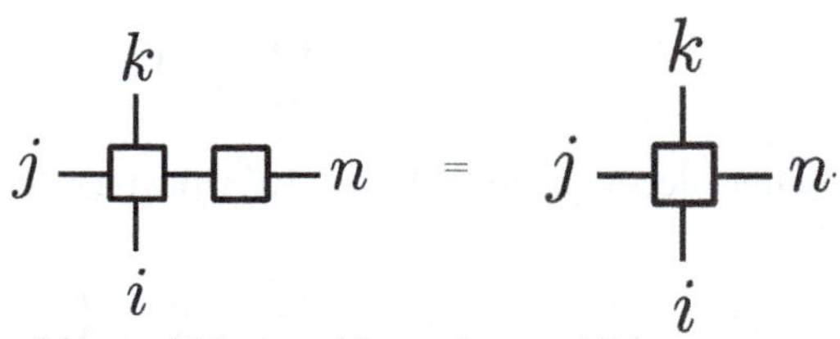

(2.100)

A TN with n unconnected legs will be used interchangeably to represent an abstract tensor and the object it returns given n arguments, depending on whether the legs are labelled or not. For example, the components of the 2-leg tensor B are $[B]_{mn}$ which is represented diagrammatically by Eq. (2.99), while the abstract tensor B itself would be represented diagrammatically as

$B =$ 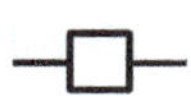
(2.101)

In general each leg in a diagram corresponds to a specific index of the tensor being represented, such that the omission of labels on legs like this could lead to ambiguity as to which leg corresponds to which index. For the purposes of this section this potential ambiguity is unimportant and the correct identification of legs with indices will always be clear from the context.

2.2.2 Matrix Products

Suppose that a 4-leg tensor, A, returns a matrix when given arguments. Further suppose that it is possible to construct this matrix through multiplication of four matrices which are themselves returned from a 1-leg tensor, C, such that

$$[A]_{ijkl} = [C]_i[C]_j[C]_k[C]_l, \tag{2.102}$$

where $[C]_i$ is the matrix returned by passing the argument i to C and multiplication such as $[C]_i[C]_j$ corresponds to standard matrix multiplication. Then $[A]_{ijkl}$ is said to admit a matrix product decomposition. A matrix is itself a tensor, since it can be considered as a function of two arguments, α and β, which returns its scalar components, i.e., $[C]_i$ is a tensor with components $[C]_{i\alpha\beta} \equiv [[C]_i]_{\alpha\beta}$. Then multiplication between matrices in a product is a contraction and can be represented by a TN,

$$\sum_\beta [C]_{i\alpha\beta}[C]_{j\beta\gamma} =$$

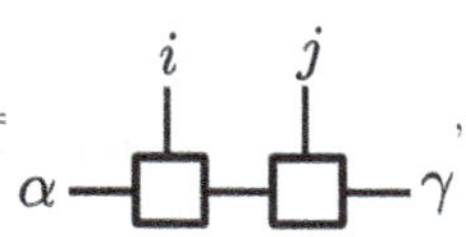

, (2.103)

or alternatively

$$[C]_i[C]_j =$$

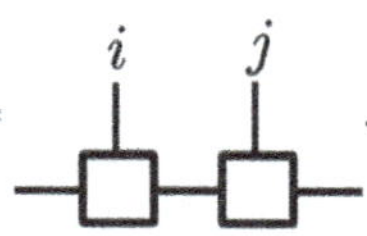

. (2.104)

The matrix product decomposition, Eq. (2.102), then has the TN representation

$$[A]_{ijkl} =$$

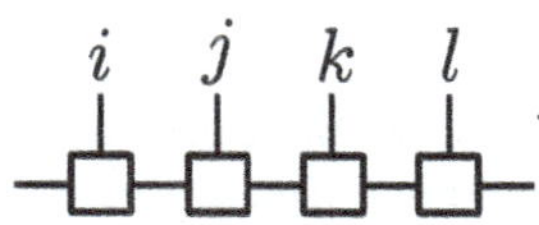

. (2.105)

The matrices that make up $[A]_{ijkl}$ need not be square matrices. That is, in Eq. (2.103) the arguments α, β, and γ need to take value from the same number of scalar values. The number of scalar values that an argument representing a single leg of a matrix in a product decomposition can take is called a bond dimension, and the single line that connects two adjacent matrices in a TN diagram of a matrix product is called a bond.

Suppose that the 4-leg tensor A returns a scalar value, not a matrix. It is still possible that a matrix product decomposition of the form Eq. (2.102) can be found provided that $[C]_i$ and $[C]_l$ are vectors instead of matrices. The TN representation of the scalar $[A]_{ijkl}$ would then be

$[A]_{ijkl} =$

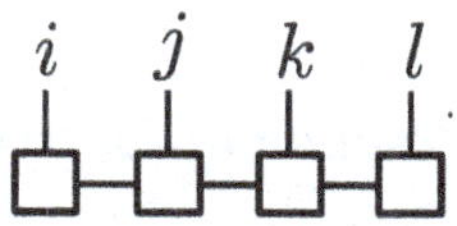

(2.106)

A matrix product decomposition of this form is called a matrix product state (MPS).

Consider an 8-leg tensor, B, that returns a scalar. For convenience, both superscripts and subscripts shall be used to represent arguments passed to this tensor $[B]_{ijkl}^{i'j'k'l'} \equiv [B]_{ijkli'j'k'l'}$. Suppose that it is possible to make the matrix product decomposition

$$[B]_{ijkl}^{i'j'k'l'} = [C]_i^{i'}[C]_j^{j'}[C]_k^{k'}[C]_l^{l'}, \qquad (2.107)$$

where $[C]_i^{i'}$, $[C]_l^{l'}$ are vectors and $[C]_j^{j'}$, $[C]_k^{k'}$ are matrices. Then $[B]_{ijkl}^{i'j'k'l'}$ can be represented as TN,

$[B]_{ijkl}^{i'j'k'l'} =$

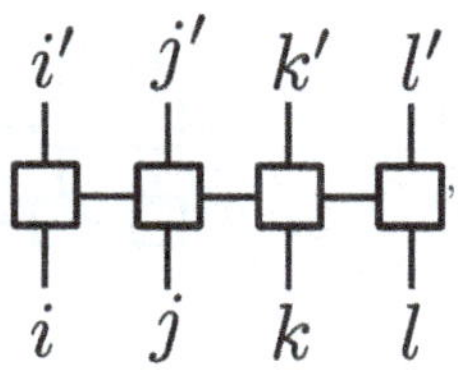

(2.108)

called a matrix product operator (MPO).

The main advantage gained by using matrix product decompositions is in reducing storage requirements for performing numerical calculations. If the arguments of the 4-leg tensor, A, can take d possible values then there are d^4 scalars that need to be stored to perform manipulations on A. Assuming that the matrices in the product that makes up the MPS representation of A, Eq. (2.106), are square matrices of dimension λ, then A has bond dimension λ. Then rather than storing A numerically as an object made up of d^4 scalars, it can be stored as the matrices and vectors which make up its matrix product decomposition which are collectively made up of $2d(\lambda + \lambda^2)$ scalars. Similarly, the MPO representation of B, Eq. (2.108), requires storage of $2d^2(\lambda + \lambda^2)$ scalars rather than d^8. Whether or not the matrix product representation of A and B do indeed admit a lower storage requirements depends on the bond dimension.

In the above example of an MPS decomposition of a 4-leg tensor, a bond dimension of $\lambda \sim d$ will always result in an MPS whose size is $\mathcal{O}(d^3)$, compared to size of the full tensor, $\mathcal{O}(d^4)$. For a general n-leg tensor that has an MPS representation in terms of a product of $n - 2$ matrices and 2 vectors with bond dimension λ, the number of scalars required for storage of the MPS is $d(2\lambda + (n - 2)\lambda^2)$, compared to the d^n scalars that make up the full n-leg tensor. Thus in general, and provided λ scales at most polynomially with n, an MPS representation of a given tensor reduces the scaling of its storage on the number of legs from exponential to polynomial. The problem of finding the most efficient representation of an n-leg tensor with a matrix product decomposition is then to find such a decomposition with small bond dimensions.

2.2.3 *Efficient Contraction of Matrix Products*

Many problems in many fields of science reduce to the problem of multiplying matrices into vectors. When the matrices and vectors in question have large dimension, often the case in many-body physics, finding MPO and MPS representations respectively can allow for more efficient numerical implementation of their multiplication. The 8-leg tensor B, whose MPO representation is given by Eq. (2.108), could for instance represent a $d^4 \times d^4$ matrix, while the 4-leg tensor A, whose MPS representation is given by Eq. (2.106), could represent a d^4 dimensional vector. Thus the multiplication of B into A to produce another d^4 dimensional vector A' can be written as a contraction between the two,

$$[A']_{ijkl} = \sum_{i'j'k'l'} [B]^{i'j'k'l'}_{ijkl} [A]_{i'j'k'l'} =$$

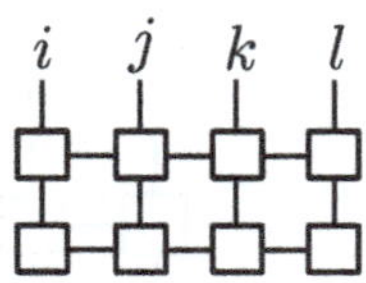

$$=$$

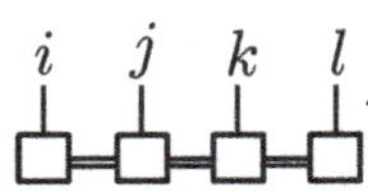

. (2.109)

The new vector, A', is also decomposable into an MPS but with a larger bond dimension, λ^2 if the bond dimensions of A and B are both λ. This is shown by the third equality in Eq. (2.109), in which two bonds of dimensions λ are equivalently represented as a single bond of dimension λ^2.

A typical problem that is solvable numerically using matrix multiplication is to find steady state vectors of a given matrix that is too large to directly diagonalise. Finding such vectors requires repeated application of the matrix B to a vector which is stored, and it is easy to see that if this vector is stored as a MPS after each contraction with the MPO B, the bond dimension of the MPS, and hence its storage requirements, will grow exponentially. If an MPS/MPO representation of the problem was not used the storage of would be constant, so using an MPS representation for A in this case appears rather to be a hindrance. There are many cases however where it is known that the matrices that make up an MPS are known to contain much redundant information. The classic example of this situation is when an MPS represents a many-body quantum state, each external leg of the MPS representing a single degree of freedom, in which there is only low entanglement between different degrees of freedom [110]. The standard tool to numerically identify only the relevant components of an MPS that need to be stored to faithfully reproduce the full tensor it represents is the singular value decomposition (SVD) [111, 112]. An SVD of a $d \times d'$ matrix with components $[M]_{\alpha\beta}$ takes the form

$$[M]_{\alpha\beta} = \sum_{\gamma=1}^{d} \sum_{\gamma'=1}^{d'} [U]_{\alpha\gamma} [S]_{\gamma\gamma'} [V]_{\gamma'\beta}, \tag{2.110}$$

where $[U]_{\alpha\gamma}$ are the components of a $d \times d$ unitary matrix U, $[V]_{\alpha\gamma}$ those of a $d' \times d'$ unitary matrix V, and $[S]_{\gamma\gamma'}$ are the components of $d \times d'$ diagonal matrix whose nonzero entries are the singular values of the matrix M. A more efficient representation of M can be found by discarding those entries of S in Eq. (2.110), the singular values, that are small, resulting in a new matrix, $\tilde{M} \approx M$. How small a singular value should be in order that its neglect makes the approximation $\tilde{M} \approx M$ a good one generally depends on the specific problem at hand, but under quite general conditions truncating singular values results in the best possible low rank approximation of a given matrix [113]. Rather than just setting singular values to zero it is more practical to instead discard entirely those rows and columns of S that contain the small singular values, $S \rightarrow \tilde{S}$, along with the corresponding columns of U, $U \rightarrow \tilde{U}$, and rows of V, $V \rightarrow \tilde{V}$. If there are d_c singular values remaining after discarding small ones then $\tilde{S}$ is a $d_c \times d_c$ square matrix, $\tilde{U}$ a $d \times d_c$ rectangular matrix, and $\tilde{V}$ a $d_c \times d'$ rectangular matrix. Overall then the approximation resulting from the neglect of all but the d_c largest singular values reads $M \approx \tilde{M} = \tilde{U}\tilde{S}\tilde{V}$.

When using SVDs to reduce the size of an MPS the neglect of small singular values can be achieved by inserting approximate identity matrices in between each matrix in the product. Consider two matrices that appear in an MPS whose bond dimension is to be truncated. Each of the tensors that represents a single site in an MPS has 3 legs, an the first step in implementing an SVD is to gather two of the legs of the right hand site into a single leg with larger dimension

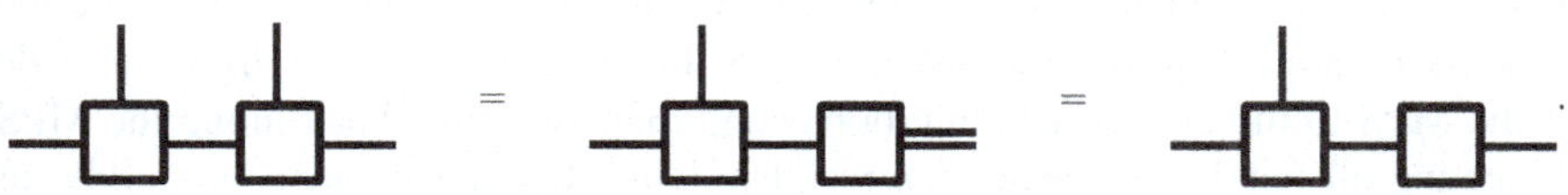

(2.111)

The second equality indicates that two legs are being represented as one. If the bond dimension is λ and the dimension of the free legs of the MPS are d then the dimension of the rightmost leg in the second equality is $d\lambda$. An SVD is then made upon the $\lambda \times d\lambda$ matrix in the second equality and some number, λ_c, of these singular values are deemed large enough to be retained. The $\lambda \times \lambda_c$ matrix $\tilde{U}$ is then extracted and used to construct an approximate $\lambda \times \lambda$ identity matrix,

$$I \approx \tilde{U}\tilde{U}^{\dagger}. \tag{2.112}$$

which becomes an equality only if no columns of U are discarded such that $\lambda = \lambda_c$ and $U = \tilde{U}$. It should be noted that at this stage no approximation has yet been made upon the MPS. The SVD was performed solely to obtain $\tilde{U}$ as a separate matrix. The approximation made upon the MPS is to insert the approximate identity, Eq. (2.112), between the two sites of the MPS,

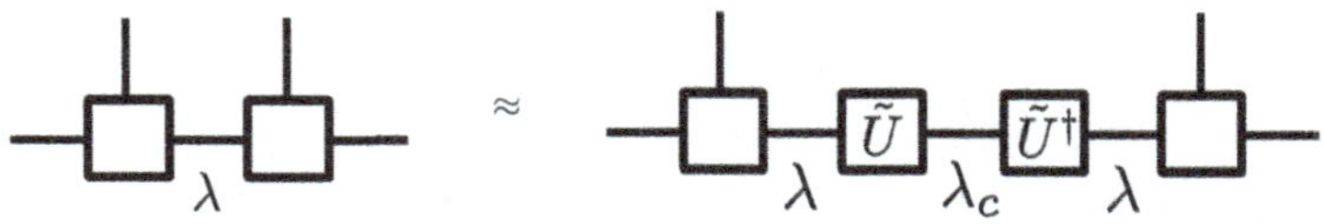

(2.113)

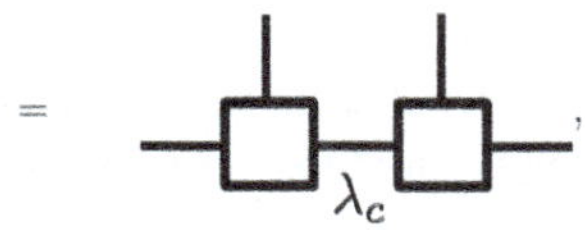

(2.114)

where the bond dimensions and matrices $\tilde{U}$, $\tilde{U}^\dagger$ have been marked for clarity. In going from the first to second line of Eq. (2.113) the left and right MPS sites have been redefined to contain $\tilde{U}$ and $\tilde{U}^\dagger$ respectively. Since the bond between the $\tilde{U}$ and $\tilde{U}^\dagger$ is $\lambda_c < \lambda$ this reduces the size of both sites.

Say that the MPS A' in Eq. (2.109) is known to contain redundant information and its bond dimension, λ^2, is reducible by performing SVDs. One way to reduce this bond dimension would be to use the procedure described above that culminated in Eq. (2.113) upon each individual bond of A' after all contractions between A and B have been performed. The more efficient way to do this is to perform the SVDs and truncations as A and B are being contracted together site-by-site. Efficiency can also be affected by what order the sites of the MPS and MPO are contracted together while truncating bond dimensions. It is also common to perform truncations of bond dimensions more than once across the MPS. For instance, moving from one side of the MPS to the other truncating bonds and then sweeping back along the MPS truncating all bonds once more can result in bond dimensions near the middle of the MPS larger than those at the edges, though an optimal order for performing truncations is generally situation dependent.

2.2.4 Summary

This section has provided a brief exposition of the the core ideas behind TNs and those parts of the theory that will be relevant in following discussions. Using TNs is an attractive way to tackle complex problems, since they provide a means to both break down a such problems into smaller parts and to represent them with simple diagrams. The standard and general notation used for writing down tensor networks diagrammatically was introduced in Sect. 2.2.1. In Sect. 2.2.2 a particular class of TN was introduced, the matrix product. Representation of tensors as matrix products potentially allows their storage and numerical manipulation at much smaller computational cost. Essentially all problems in linear algebra are solvable through matrix-vector multiplication and it was shown how it is possible to represent such multiplication as contraction between a MPO and a MPS. If MPOs are repeatedly contracted into a MPS the storage of the MPS grows exponentially, negating the use of an MPS representation to improve storage requirements. In Sect. 2.2.3 it was shown how singular value decompositions can be used in conjunction with a MPO/MPS representation of to maintain a small sized MPS. This involves discarding components of the MPS that are irrelevant and represents a controlled numerical approximation.

References

1. Breuer H-P, Petruccione F (2002) The Theory of Open Quantum Systems. Oxford University Press
2. Weiss U (2012) Quantum Dissipative Systems. World Scientific
3. Nakajima S (1958) On quantum theory of transport phenomena: steady diffusion. Prog Theor Phys 20(6):948
4. Zwanzig R (1960) Ensemble method in the theory of irreversibility. J Chem Phys 33(5):1338
5. Ford GW, O'Connell RF (1996) There is no quantum regression theorem. Phys Rev Lett 77:798
6. Gardiner C, Zoller P (2004) Quantum Noise: A Handbook of Markovian and Non-Markovian Quantum Stochastic Methods with Applications to Quantum Optics. Springer Series in Synergetics. Springer, Berlin
7. Breuer H-P, Petruccione F (2001) Destruction of quantum coherence through emission of bremsstrahlung. Phys Rev A 63:032102
8. Feynman RP, Hibbs AR (1965) Quantum Mechanics and Path Integrals. McGraw-Hill, New York
9. Feynman RP, Vernon FL (1963) The theory of a general quantum system interacting with a linear dissipative system. Ann Phys 24:118
10. Ingold G-L (2002) Path integrals and their application to dissipative quantum systems. In Coherent Evolution in Noisy Environments. Springer, Berlin Heidelberg, p 1
11. de Vega I, Alonso D (2017) Dynamics of non-Markovian open quantum systems. Rev Mod Phys 89:015001
12. Breuer H-P, Laine E-M, Piilo J, Vacchini B (2016) Colloquium: non-Markovian dynamics in open quantum systems. Rev ModPhys 88:021002
13. Ferialdi L (2016) Exact closed master equation for gaussian non-Markovian dynamics. Phys Rev Lett 116:120402
14. Breuer H-P, Ma A, Petruccione F (2004) Time-Local Master Equations: Influence Functional and Cumulant Expansion. Springer, US, Boston, MA
15. Dyson FJ (1949) The radiation theories of Tomonaga, Schwinger, and Feynman. Phys Rev 75(3):486
16. Rivas Á, Huelga SF (2012) Open Quantum Systems. Springer, Berlin
17. Diósi L, Ferialdi L (2014) General non-Markovian structure of gaussian master and stochastic schrödinger equations. Phys Rev Lett 113:200403
18. Caldeira AO, Leggett AJ (1983) Path integral approach to quantum Brownian motion. Phys A 121(3):587
19. Caldeira AO, Leggett AJ (1983) Quantum tunnelling in a dissipative system. Ann Phys (N.Y.) 149(2):374

20. Caldeira AO, Leggett AJ (1981) Influence of dissipation on quantum tunneling in macroscopic systems. Phys Rev Lett 46:211
21. Mahan GD (2000) Many Particle Physics, 3rd edn. Springer, US
22. Walls DF, Milburn GJ (2007) Quantum Optics. Springer, Berlin
23. Leggett AJ, Chakravarty S, Dorsey AT, Fisher MPA, Garg A, Zwerger W (1987) Dynamics of the dissipative two-state system. Rev Mod Phys 59(1):1
24. Kok P, Lovett BW (2010) Introduction to optical quantum information processing. Cambridge University Press, Cambridge
25. DiVincenzo D, Terhal B (1998) Decoherence: the obstacle to quantum computation. Phys World 11(3):53
26. Nazir A, McCutcheon DPS (2016) Modelling exciton-phonon interactions in optically driven quantum dots. J Phys Condens Matter 28(10):103002
27. Ramsay AJ, Godden TM, Boyle SJ, Gauger EM, Nazir A, Lovett BW, Fox AM, Skolnick MS (2010) Phonon-induced Rabi-frequency renormalization of optically driven single InGaAs/GaAs quantum dots. Phys Rev Lett 105:177402
28. Ramsay AJ, Gopal AV, Gauger EM, Nazir A, Lovett BW, Fox AM, Skolnick MS (2010) Damping of exciton Rabi rotations by acoustic phonons in optically excited InGaAs/GaAs quantum dots. Phys Rev Lett 104:017402
29. Unitt DC, Bennett AJ, Atkinson P, Cooper K, See P, Gevaux D, Ward MB, Stevenson RM, Ritchie DA, Shields AJ (2005) Quantum dots as single-photon sources for quantum information processing. J Opt B: Quantum Semiclass Opt 7(7):S129
30. Michler P (2000) A quantum dot single-photon turnstile device. Science 290(5500):2282
31. Lovett BW, Reina JH, Nazir A, Briggs AI (2003) Optical schemes for quantum computation in quantum dot molecules. Phys Rev B 68(20):205319
32. Lv D, An S, Liu Z, Zhang J-N, Pedernales JS, Lamata L, Solano E, Kim K (2018) Quantum simulation of the quantum Rabi model in a trapped ion. Phys Rev X 8(2):021027
33. Moya-Cessa HM (2016) Fast quantum Rabi model with trapped ions. Sci Rep 6:38961
34. Braumüller J, Marthaler M, Schneider A, Stehli A, Rotzinger H, Weides M, Ustinov AV (2017) Analog quantum simulation of the Rabi model in the ultra-strong coupling regime. Nat Commun 8:779
35. Forn-Díaz P, Romero G, Harmans CJPM, Solano E, Mooij JE (2016) Broken selection rule in the quantum Rabi model. Sci Rep 6:26720
36. Roy C, Hughes S (2012) Polaron master equation theory of the quantum-dot Mollow triplet in a semiconductor cavity-QED system. Phys Rev B 85(11):115309
37. Fischer KA, Hanschke L, Kremser M, Finley JJ, Müller K, Vučković J (2017) Pulsed Rabi oscillations in quantum two-level systems: beyond the area theorem. Quantum Sci Technol 3(1):014006
38. Lindblad G (1976) On the generators of quantum dynamical semigroups. Commun Math Phys 48(2):119
39. Gorini V (1976) Completely positive dynamical semigroups of n-level systems. J Math Phys 17(5):821
40. Haikka P, Johnson TH, Maniscalco S (2013) Non-Markovianity of local dephasing channels and time-invariant discord. Phys Rev A 87(1):010103
41. Redfield AG (1965) The theory of relaxation processes. In Advances in Magnetic Resonance. Elsevier, p 1
42. Mollow BR (1969) Power spectrum of light scattered by two-level systems. Phys Rev 188(5):1969
43. Liu Z-D, Lyyra H, Sun Y-N, Liu B-H, Li C-F, Guo G-C, Maniscalco S, Piilo J (2018) Experimental implementation of fully controlled dephasing dynamics and synthetic spectral densities. Nat Commun 9:3453
44. Ros V, Müller M, Scardicchio A (2015) Integrals of motion in the many-body localized phase. Nucl Phys B 891:420
45. Ali MM, Lo P-Y, Tu MW-Y, Zhang W-M (2015) Non-Markovianity measure using two-time correlation functions. Phys Rev A 92:062306

46. Milz S, Kim MS, Pollock FA, Modi K (2019) CP divisibility does not mean Markovianity. arXiv:1901.05223
47. Breuer H-P, Laine E-M, Piilo J (2009) Measure for the degree of non-Markovian behavior of quantum processes in open systems. Phys Rev Lett 103:210401
48. Laine E-M, Piilo J, Breuer H-P (2010) Measure for the non-Markovianity of quantum processes. Phys Rev A 81:062115
49. Rivas Á, Huelga SF, Plenio MB (2010) Entanglement and non-Markovianity of quantum evolutions. Phys Rev Lett 105:050403
50. Chitambar E, Hsieh M-H (2016) Relating the resource theories of entanglement and quantum coherence. Phys Rev Lett 117(2):020402
51. Horodecki R, Horodecki P, Horodecki M, Horodecki K (2009) Quantum entanglement. Rev Mod Phys 81(2):865
52. Coecke B, Fritz T, Spekkens RW (2016) A mathematical theory of resources. Inf Comput 250:59
53. Bylicka B, Chruściński D, Maniscalco S (2014) Non-Markovianity and reservoir memory of quantum channels: a quantum information theory perspective. Sci Rep 4:5720
54. Sakuldee F, Milz S, Pollock FA, Modi K (2018) Non-Markovian quantum control as coherent stochastic trajectories. J Phys A: Math Theor 51(41):414014
55. Kitajima S, Ban M, Shibata F (2013) A solvable dissipative jaynes-cummings model with initial correlation. J Phys B: At Mol Opt Phys 46(22):224004
56. Vacchini B, Breuer H-P (2010) Exact master equations for the non-Markovian decay of a qubit. Phys Rev A 81:042103
57. Eastham PR, Kirton P, Cammack HM, Lovett BW, Keeling J (2016) Bath-induced coherence and the secular approximation. Phys Rev A 94:012110
58. Hu BL, Paz JP, Zhang Y (1992) Quantum Brownian motion in a general environment: exact master equation with nonlocal dissipation and colored noise. Phys Rev D 45(8):2843
59. Halliwell JJ, Yu T (1996) Alternative derivation of the hu-paz-zhang master equation of quantum Brownian motion. Phys Rev D 53(4):2012
60. Ferialdi L (2017) Dissipation in the caldeira-leggett model. Physical Review A 95(5)
61. Würger A (1998) Strong-coupling theory for the spin-phonon model. Phys Rev B 57(1):347
62. McCutcheon DPS, Nazir A (2010) Quantum dot Rabi rotations beyond the weak exciton-phonon coupling regime. New J Phys 12(11):113042
63. McCutcheon DPS (2016) Optical signatures of non-Markovian behavior in open quantum systems. Phys Rev A 93(2):022119
64. Roy-Choudhury K, Hughes S (2015) Quantum theory of the emission spectrum from quantum dots coupled to structured photonic reservoirs and acoustic phonons. Phys Rev B 92(20):205406
65. Dekker H (1987) Noninteracting-blip approximation for a two-level system coupled to a heat bath. Phys Rev A 35:1436
66. McCutcheon DPS, Dattani NS, Gauger EM, Lovett BW, Nazir A (2011) A general approach to quantum dynamics using a variational master equation: application to phonon-damped Rabi rotations in quantum dots. Phys Rev B 84:081305
67. Pollock FA, McCutcheon DPS, Lovett BW, Gauger EM, Nazir A (2013) A multi-site variational master equation approach to dissipative energy transfer. New J Phys 15(7):075018
68. Lee CK, Moix J, Cao J (2012) Accuracy of second order perturbation theory in the polaron and variational polaron frames. J Chem Phys 136(20):204120
69. Mazzola L, Maniscalco S, Piilo J, Suominen K-A, Garraway BM (2009) Pseudomodes as an effective description of memory: non-markovian dynamics of two-state systems in structured reservoirs. Phys Rev A 80:012104
70. Garg A, Onuchic JN, Ambegaokar V (1985) Effect of friction on electron transfer in biomolecules. J Chem Phys 83(9):4491
71. Hughes KH, Christ CD, Burghardt I (2009) Effective-mode representation of non-Markovian dynamics: a hierarchical approximation of the spectral density. II. Application to environment-induced nonadiabatic dynamics. J Chem Phys 131(12):124108

72. Iles-Smith J, Lambert N, Nazir A (2014) Environmental dynamics, correlations, and the emergence of noncanonical equilibrium states in open quantum systems. Phys Rev A 90(3):032114
73. Iles-Smith J, Dijkstra AG, Lambert N, Nazir A (2016) Energy transfer in structured and unstructured environments: master equations beyond the born-markov approximations. J Chem Phys 144(4):044110
74. Strasberg P, Schaller G, Lambert N, Brandes T (2016) Nonequilibrium thermodynamics in the strong coupling and non-Markovian regime based on a reaction coordinate mapping. New J Phys 18(7):073007
75. Chin AW, Rivas Á, Huelga SF, Plenio MB (2010) Exact mapping between system-reservoir quantum models and semi-infinite discrete chains using orthogonal polynomials. J Math Phys 51(9):092109
76. Woods MP, Groux R, Chin AW, Huelga SF, Plenio MB (2014) Mappings of open quantum systems onto chain representations and Markovian embeddings. J Math Phys 55(3):032101
77. Woods MP, Cramer M, Plenio MB (2015) Simulating bosonic baths with error bars. Phys Rev Lett 115(13):130401
78. Prior J, Chin AW, Huelga SF, Plenio MB (2010) Efficient simulation of strong system-environment interactions. Phys Rev Lett 105(5):050404
79. Chin AW, Prior J, Rosenbach R, Caycedo-Soler F, Huelga SF, Plenio MB (2013) The role of non-equilibrium vibrational structures in electronic coherence and recoherence in pigment-protein complexes. Nat Phys 9(2):113
80. Makri N, Makarov DE (1995) Tensor propagator for iterative quantum time evolution of reduced density matrices. I. Theory. J Chem Phys 102(11):4600
81. Makri N, Makarov DE (1995) Tensor propagator for iterative quantum time evolution of reduced density matrices. II. Numerical methodology. J Chem Phys 102(11):4611
82. Shao J, Makri N (2002) Iterative path integral formulation of equilibrium correlation functions for quantum dissipative systems. J Chem Phys 116(2):507
83. Cosacchi M, Cygorek M, Ungar F, Barth AM, Vagov A, Axt VM (2018) Path-integral approach for nonequilibrium multitime correlation functions of open quantum systems coupled to Markovian and non-Markovian environments. Phys Rev B 98(12):125302
84. Sim E (2001) Quantum dynamics for a system coupled to slow baths: on-the-fly filtered propagator method. J Chem Phys 115(10):4450
85. Sim E, Makri N (1997) Filtered propagator functional for iterative dynamics of quantum dissipative systems. Comput Phys Commun 99(2–3):335
86. Makri N (2017) Iterative blip-summed path integral for quantum dynamics in strongly dissipative environments. J Chem Phys 146(13):134101
87. Segal D, Millis AJ, Reichman DR (2010) Numerically exact path-integral simulation of nonequilibrium quantum transport and dissipation. Phys Rev B 82(20):205323
88. Barth AM, Vagov A, Axt VM (2016) Path-integral description of combined Hamiltonian and non-Hamiltonian dynamics in quantum dissipative systems. Phys Rev B 94(12):125439
89. Cygorek M, Barth AM, Ungar F, Vagov A, Axt VM (2017) Nonlinear cavity feeding and unconventional photon statistics in solid-state cavity QED revealed by many-level real-time path-integral calculations. Phys Rev B 96(20):201201
90. Makri N (2012) Path integral renormalization for quantum dissipative dynamics with multiple timescales. Mol Phys 110(9–10):1001
91. Lambert R, Makri N (2012) Memory propagator matrix for long-time dissipative charge transfer dynamics. Mol Phys 110(15–16):1967
92. Walters PL, Banerjee T, Makri N (2015) On iterative path integral calculations for a system interacting with a shifted dissipative bath. J Chem Phys 143(7):074112
93. Ilk G, Makri N (1994) Real time path integral methods for a system coupled to an anharmonic bath. J Chem Phys 101(8):6708
94. Cygorek M, Ungar F, Seidelmann T, Barth AM, Vagov A, Axt VM, Kuhn T (2018) Comparison of different concurrences characterizing photon pairs generated in the biexciton cascade in quantum dots coupled to microcavities. Phys Rev B 98(4):045303

95. Thorwart M, Reimann P, Hänggi P (2000) Iterative algorithm versus analytic solutions of the parametrically driven dissipative quantum harmonic oscillator. Phys Rev E 62(4):5808
96. Nalbach P, Thorwart M (2010) Ultraslow quantum dynamics in a sub-ohmic heat bath. Phys Rev B 81(5):054308
97. Nalbach P, Ishizaki A, Fleming GR, Thorwart M (2011) Iterative path-integral algorithm versus cumulant time-nonlocal master equation approach for dissipative biomolecular exciton transport. New J Phys 13(6):063040
98. Golosov AA, Friesner RA, Pechukas P (1999) Efficient memory equation algorithm for reduced dynamics in spin-boson models. J Chem Phys 110(1):138
99. Cerrillo J, Cao J (2014) Non-Markovian dynamical maps: numerical processing of open quantum trajectories. Phys Rev Lett 112(11):110401
100. Buser M, Cerrillo J, Schaller G, Cao J (2017) Initial system-environment correlations via the transfer-tensor method. Phys Rev A 96(6):062122
101. Tanimura Y, Kubo R (1989) Time evolution of a quantum system in contact with a nearly gaussian-markoffian noise bath. J Phys Soc Jpn 58(1):101
102. Tanimura Y (2014) Reduced hierarchical equations of motion in real and imaginary time: correlated initial states and thermodynamic quantities. J Chem Phys 141(4):044114
103. Tang Z, Ouyang X, Gong Z, Wang H, Wu J (2015) Extended hierarchy equation of motion for the spin-boson model. J Chem Phys 143(22):224112
104. Ma J, Sun Z, Wang X, Nori F (2012) Entanglement dynamics of two qubits in a common bath. Phys Rev A 85(6):062323
105. Duan C, Tang Z, Cao J, Wu J (2017) Zero-temperature localization in a sub-ohmic spin-boson model investigated by an extended hierarchy equation of motion. Phys Rev B 95(21):214308
106. Goan H-S, Jian C-C, Chen P-W (2010) Non-Markovian finite-temperature two-time correlation functions of system operators of a pure-dephasing model. Phys Rev A 82(1):012111
107. Guarnieri G, Smirne A, Vacchini B (2014) Quantum regression theorem and non-Markovianity of quantum dynamics. Phys Rev A 90:022110
108. Orús R (2014) A practical introduction to tensor networks: matrix product states and projected entangled pair states. Ann Phys (N.Y.) 349(Supplement C):117
109. Schuch N, Pérez-García D, Cirac I (2011) Classifying quantum phases using matrix product states and projected entangled pair states. Phys Rev B 84(16):165139
110. Verstraete F, Murg V, Cirac JI (2008) Matrix product states, projected entangled pair states, and variational renormalization group methods for quantum spin systems. Adv Phys 57(2):143–224
111. Oseledets IV (2011) Tensor-train decomposition. SIAM J Sci Comput 33(5):2295
112. Daley AJ, Kollath C, Schollwöck U, Vidal G (2004) Time-dependent density-matrix renormalization-group using adaptive effective Hilbert spaces. J Stat Mech Theory Exp 2004(04):P04005
113. Eckart C, Young G (1936) The approximation of one matrix by another of lower rank. Psychometrika 1(3):211–218

Chapter 3
Method

Fundamentally, all non-unitary behaviour of an open quantum system, Markovian or non-Markovian, mathematically originates from time-nonlocality present in its Liouvillian. Despite the unavoidable emergence of time-nonlocal Liouvillians in modelling open systems, there is no known practical and general way of using such Liouvillians to calculate observables exactly, even if a closed form of the Liouvillian can be found. It is not even known what the most general closed form of time-nonlocal Liouvillian is that guarantees physicality of an open system. However, if an arbitrary Liouvillian is only time-nonlocal due to a single time integral over its argument functional then it must be of the same form as a Liouvillian that describes non-abelian interaction with some HB [1]. As a conceptual barrier standing between closed form models and calculation of observables time-nonlocality is not well understood. The problem can be summed up by the statement that the time-ordering operator, $\overleftarrow{\mathrm{T}}$, is defined by its action only on products of time-local operators, and so appears to be ill defined when confronted with time-nonlocal arguments such as an IF [2]. In this chapter I propose a direct and general solution to this problem that, in principle, allows for numerical calculation of arbitrary observables of arbitrary time-nonlocal open quantum systems.

In Sect. 3.1 it is shown how time ordering of arbitrary functionals can be acheived by contracting a TN, the only assumption required being that it is possible to discretise time onto a finite set of points. This result is significant since it elevates the notion of time ordering from a formal mathematical bookkeeping device to a mathematical operation that a computer can be programmed to carry out. The properties of any given open quantum system can be calculated through time ordering of some expression that involves an IF, and in Sect. 3.2 it is shown how knowledge of the closed form for an IF can be used to reveal a simple structure to the corresponding TN representation. In particular, the TN representing the time ordering of a Gaus-

A. Strathearn, *Modelling Non-Markovian Quantum Systems Using Tensor Networks*, Springer Theses, https://doi.org/10.1007/978-3-030-54975-6_3

sian IF is given explicitly and it is shown how the Redfield approximation discussed in Sect. 2.1.2.2 can be generalised such that the time ordering of arbitrary IFs can represented by such a TN. It is then shown in Sect. 3.3 how the TN for a Gaussian IF can be put in the form of a series of MPOs. The calculation of observables of an open quantum system can then be carried out by iterative contraction between an MPS and an MPO, which can be made efficient using the techniques discussed in Sect. 2.2. By exploiting the fact that time-nonlocality in an open system is typically finite-ranged, a further approximation, the finite memory approximation, can be made that reduces the size of the MPOs and MPSs. This improves efficiency even further and is the subject of Sect. 3.4. The finite memory approximation, like the Redfield approximation, is long established, but the consequences of making this approximation have received little analytical attention. In Sect. 3.5 the error induced by making the finite memory approximation in an exactly solvable pure dephasing is found to be rather sensitive to the details of the model. An improvement to this approximation, inspired by the Redfield approximation, is then suggested. Finally, the proposed methods for efficiently carrying out the time ordering of an IF are demonstrated in practice in Sect. 3.6. The model considered is of an open system with a Feynman-Vernon IF, whose path integral representation is known to be numerically tractable using the finite memory approximation [3]. It is shown that the TN representation of the problem provides significant improvement in efficiency compared to established techniques. Further numerical evidence is provided that shows the improved finite memory approximation suggested in Sect. 3.5 can outperform the established finite memory approximation discussed in Sect. 3.4.

3.1 Time Ordering as a Tensor Network

Consider a superoperator $\mathcal{Q}(t)$ found by application of time ordering, $\overleftarrow{\mathrm{T}}$, to some arbitrary functional $f[t] \equiv f[x]_0^t$ whose argument is an operator valued function $x(t)$,

$$\mathcal{Q}(t) = \overleftarrow{\mathrm{T}}\, f[t]. \tag{3.1}$$

It is assumed that $x(t)$ is a $d_s \times d_s$ operator that acts upon vectors in a d_s-dimensional vector space, $\mathcal{H}$. Then it is possible to represent superoperators, such as $x^-(t)$, $x^+(t)$, or $\mathcal{Q}(t)$, as $d \times d$ operators acting upon vectors in a d-dimensional vector space $\mathcal{G} = \mathcal{H} \otimes \mathcal{H}$, where $d = d_s^2$ [4, 5]. Within this formalism a $d_s \times d_s$ density matrix is mapped onto a d-dimensional vector, an element of $\mathcal{G}$, so that expressions such as $x^-(t)\rho(t)$ should be read as multiplication of a $d \times d$ matrix into a d-dimensional vector.

Operators in $\mathcal{G}$ can always be decomposed onto d^2 basis operators, $\hat{e}_\alpha^\beta = |\alpha\rangle\langle\beta|$, that satisfy,

$$\hat{e}_\alpha^\beta \hat{e}_{\alpha'}^{\beta'} = \hat{e}_\alpha^{\beta'} \delta_{\alpha'}^{\beta}, \tag{3.2}$$

where $|\alpha\rangle$ is a d-dimensional vector. The convention of implied summation when an upper and lower index are repeated is now adopted. Then orthogonality of basis operators can be expressed as

$$\mathrm{Tr}[\hat{e}^{\gamma}_{\alpha}\hat{e}^{\beta}_{\gamma}] = \delta^{\beta}_{\alpha}, \tag{3.3}$$

and any given operator, O, in $\mathcal{G}$ can be decomposed onto basis operators in the following way,

$$O = [O]^{\alpha}_{\beta}\hat{e}^{\beta}_{\alpha}, \tag{3.4}$$

where $[O]^{\alpha}_{\beta} = \langle\alpha|\, O\, |\beta\rangle$ are the d^2 components of O.

To find a TN representation of Eq. (3.1), the continuous time variable t is approximated as only taking values from some finite set of times $\{t\} \equiv \{t_n, t_{n-1}, \ldots, t_k, \ldots, t_1\}$ with $1 \leq k \leq n$ and

$$t_k - t_{k-1} = \Delta > 0, \tag{3.5}$$

for some Δ. The functional $f[t]$ is then no longer a functional of a continuous operator valued function, $x(t)$, but is rather a function of n operators $\{x(t_n), x(t_{n-1}), \ldots, x(t_k), \ldots, x(t_1)\}$

$$f[t] \approx f(t_n, t_{n-1}, \ldots, t_k, \ldots, t_1) \equiv f\{t\}. \tag{3.6}$$

Assuming for simplicity that operators in $\mathcal{H}$ are bosonic in nature it is straightforward to verify the property

$$\overleftarrow{\mathrm{T}}[O(t_k), O'(t_{k'})] = \delta_{kk'}[O(t_k), O'(t_{k'})], \tag{3.7}$$

where $[O(t_k), O'(t_{k'})]$ denotes the commutator between two operators $O(t_k)$ and $O'(t_{k'})$. If operators in $\mathcal{H}$ are fermionic in nature then the commutator in this relation would be replaced by an anticommutator. Thus, before $\overleftarrow{\mathrm{T}}$ is applied to $f\{t\}$, operators labeled with distinct time variables can be treated as operators acting in distinct vector spaces, since they always commute (or anticommute) with each other. That is, all operators labeled with the time variable t_k, e.g. $O(t_k)$, act in their own vector space $\mathcal{G}(t_k)$ which is identical in structure to $\mathcal{G}$. The function $f\{t\}$ can then be treated as a single operator that acts in the product space $\mathcal{G}_\Delta = \mathcal{G}(t_n) \otimes \mathcal{G}(t_{n-1}) \otimes \cdots \otimes \mathcal{G}(t_1)$. The time ordering operator is then a map, $\overleftarrow{\mathrm{T}} : \mathcal{G}_\Delta \rightarrow \mathcal{G}$. Such a construction is a key mathematical tool in Feynman operator calculus [6–8], which provides the link between the path ordering and path integral approaches to quantum mechanics.

There are d^{2n} basis operators of $\mathcal{G}_\Delta$, that can be constructed by taking the tensor product of basis operators of each subspace $\mathcal{G}(t_k)$, such that the operator $f\{t\}$ can be decomposed as

$$f\{t\} = \hat{e}^{\beta_n}_{\alpha_n} \otimes \hat{e}^{\beta_{n-1}}_{\alpha_{n-1}} \otimes \cdots \otimes \hat{e}^{\beta_k}_{\alpha_k} \otimes \cdots \otimes \hat{e}^{\beta_1}_{\alpha_1} [f\{t\}]^{\alpha_n, \alpha_{n-1}, \ldots, \alpha_k, \ldots, \alpha_1}_{\beta_n \beta_{n-1}, \ldots, \beta_k, \ldots, \beta_1}. \tag{3.8}$$

Time ordering only acts upon operators, meaning time ordering of $f\{t\}$ is achieved by time ordering the basis operators of the product space $\mathcal{G}_\Delta$. These basis operators have been constructed by taking tensor products of basis operators of the subspaces, $\mathcal{G}(t_k)$, in chronological order from right to left, so the action of $\overleftarrow{\mathrm{T}}$ on a given basis operator of $\mathcal{G}_\Delta$ is to replace tensor products between operators by matrix multiplication,

$$\overleftarrow{\mathrm{T}}\, \hat{e}^{\beta_n}_{\alpha_n} \otimes \hat{e}^{\beta_{n-1}}_{\alpha_{n-1}} \otimes \cdots \otimes \hat{e}^{\beta_k}_{\alpha_k} \otimes \cdots \otimes \hat{e}^{\beta_1}_{\alpha_1} = \hat{e}^{\beta_n}_{\alpha_n} \hat{e}^{\beta_{n-1}}_{\alpha_{n-1}} \ldots \hat{e}^{\beta_k}_{\alpha_k} \ldots \hat{e}^{\beta_1}_{\alpha_1} = \hat{e}^{\beta_1}_{\alpha_n} \prod_{k=1}^{n-1} \delta^{\beta_{k+1}}_{\alpha_k}. \tag{3.9}$$

The second equality follows from the fact that each subspace of $\mathcal{G}_\Delta$ is identical, spanned by identical sets of basis operators that all satisfy Eq. (3.2).

With the discretisation of time, the construction of the product space $\mathcal{G}_\Delta$, and the action of time ordering given by Eq. (3.9), the elements of the operator $\mathcal{Q}(t)$ in $\mathcal{G}$ can be written

$$[\mathcal{Q}(t_n)]^{\alpha_n}_{\beta_1} = \prod_{k=1}^{n-1} [\mathcal{T}_k]^{\beta_{k+1}}_{\alpha_k} [f\{t\}]^{\alpha_n \alpha_{n-1} \ldots \alpha_1}_{\beta_n \beta_{n-1} \ldots \beta_1}. \tag{3.10}$$

where

$$[\mathcal{T}_k]^{\beta_{k+1}}_{\alpha_k} = \delta^{\beta_{k+1}}_{\alpha_k} \tag{3.11}$$

Now, $f\{t\}$ can be considered as a $2n$-leg tensor with d^{2n} elements that has diagrammatic representation

$$[f\{t\}]^{\alpha_n \ldots \alpha_k \ldots \alpha_1}_{\beta_n \ldots \beta_k \ldots \beta_1} =$$

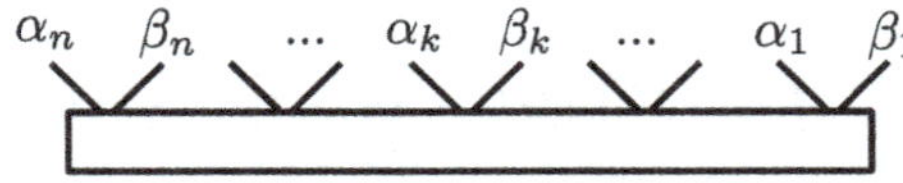

. (3.12)

Similarly each $[\mathcal{T}_k]^{\beta_{k+1}}_{\alpha_k}$ under the product in Eq. (3.10) can be considered as the d^2 elements of a 2-leg tensor $\mathcal{T}_k$,

$$[\mathcal{T}_k]^{\beta_{k+1}}_{\alpha_k} =$$

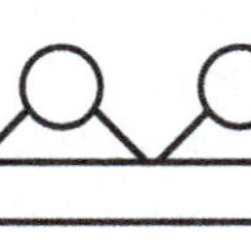

. (3.13)

Then Eq. (3.10) represents a contraction between $f\{t\}$ and the $n-1$ tensors $\mathcal{T}_k$, and can be expressed as a TN diagram,

$$[\mathcal{Q}(t_n)]^{\alpha_n}_{\beta_1} =$$

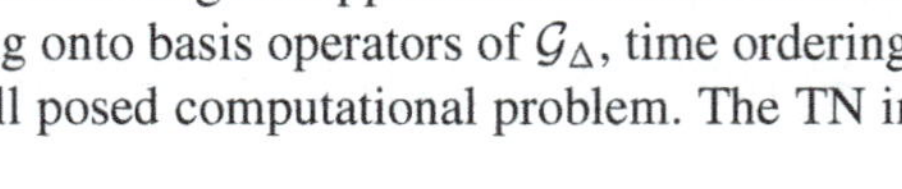

. (3.14)

Thus, provided that discretisation is a good approximation and that elements of $f\{t\}$ can be obtained by projecting onto basis operators of $\mathcal{G}_\Delta$, time ordering of arbitrary functionals is, at least, a well posed computational problem. The TN in Eq. (3.14), in the language of Feynman operator calculus, can be thought of as a diagrammatic representation of the disentangling of some operator functional $f[t]$ [8, 9].

Before continuing to consider the time ordering of IFs of open quantum systems, some useful properties of the above TN representation of time ordering are now pointed out. Consider that the discretised functional $f\{t\}$ has the form

$$f\{t\} = \prod_{k=1}^{n} O_k(t_k) f'\{t\}, \tag{3.15}$$

for some other functional $f'[t]$ whose discretisation yields $f'\{t\}$. The product of operators $O_k(t_k)$ is represented as an operator in $\mathcal{G}_\Delta$ like

$$\prod_{k=1}^{n} O_k(t_k) = O_n(t_n) \otimes O_{n-1}(t_{n-1}) \otimes \cdots \otimes O_1(t_1), \tag{3.16}$$

such that projecting $f'\{t\}$ onto basis operators of $\mathcal{G}_\Delta$ results in

$$f\{t\} = O_n(t_n)\hat{e}_{\alpha_n}^{\beta_n} \otimes O_{n-1}(t_{n-1})\hat{e}_{\alpha_{n-1}}^{\beta_{n-1}} \otimes \cdots \otimes O_1(t_1)\hat{e}_{\alpha_1}^{\beta_1} [f'\{t\}]_{\beta_n \beta_{n-1},\ldots,\beta_k,\ldots,\beta_1}^{\alpha_n,\alpha_{n-1},\ldots,\alpha_k,\ldots,\alpha_1} . \tag{3.17}$$

Then application of $\overleftarrow{\mathrm{T}}$ gives an expression identical in form to Eq. (3.10) with the same TN diagram, Eq. (3.14), but with $f \to f'$ and

$$[\mathcal{T}_k]_{\alpha_k}^{\beta_{k+1}} = [O_k(t_k)]_{\alpha_k}^{\beta_{k+1}}. \tag{3.18}$$

It might be the case that the time dependence of operators within the functional $f[t]$ is due to some unitary transformation,

$$x^{\pm}(t) = U(t) x^{\pm} U^{\dagger}(t), \tag{3.19}$$

where $x^{\pm}$ is defined as in Eq. (2.31) and

$$U(t) = \overleftarrow{\mathrm{T}}\, e^{i \int_0^t h^-(t')dt'}, \tag{3.20}$$

with $h^-(t)\rho(t) = [h(t), \rho(t)]$ for some Hermitian operator $h(t)$. Note that Eq. (3.19) is a transformation being made on a $d \times d$ operator, $x^{\pm}$, which acts in $\mathcal{G}$ using unitary operators, $U(t)$, which are also $d \times d$ operators in $\mathcal{G}$. The unitary transformation of $\rho(t)$, a d-dimensional vector in $\mathcal{G}$, would then be written $U(t)\rho(t)$. It is now possible to project operators onto a time dependent basis operators such that

$$x^{\pm}(t_k) = \hat{e}_{\alpha}^{\beta}(t_k)[x^{\pm}]_{\beta}^{\alpha} = U(t_k)\hat{e}_{\alpha}^{\beta} U^{\dagger}(t_k)[x^{\pm}]_{\beta}^{\alpha}, \tag{3.21}$$

for all t_k. The discretised functional $f\{t\}$ can then be expanded like,

$$f\{t\} = U(t_n)\hat{e}^{\beta_n}_{\alpha_n} U^\dagger(t_n) \otimes U(t_{n-1})\hat{e}^{\beta_{n-1}}_{\alpha_{n-1}} U^\dagger(t_{n-1}) \otimes \cdots \otimes U(t_1)\hat{e}^{\beta_1}_{\alpha_1} U^\dagger(t_1) \times [f\{t\}]^{\alpha_n,\alpha_{n-1},\ldots,\alpha_k,\ldots,\alpha_1}_{\beta_n\beta_{n-1},\ldots,\beta_k,\ldots,\beta_1}. \quad (3.22)$$

Application of time ordering reveals that

$$\mathcal{Q}(t_n) = U(t_n)\mathcal{Q}'(t_n)U^\dagger(t_1), \quad (3.23)$$

where $\mathcal{Q}'(t_n)$ has a TN representation the same as Eq. (3.14), but with components of $f\{t\}$ found using time dependent basis operators, $\hat{e}^{\beta}_{\alpha}(t_k)$, and with

$$[\mathcal{T}_k]^{\beta_{k+1}}_{\alpha_k} = [U^\dagger(t_{k+1})U(t_k)]^{\beta_{k+1}}_{\alpha_k} = [U^\dagger(\Delta)]^{\beta_{k+1}}_{\alpha_k}, \quad (3.24)$$

where the property $U^\dagger(t)U(t') = U^\dagger(t - t')$ has been used. If in this case $f\{t\}$ can also be factorised as in Eq. (3.15) then the TN for $\mathcal{Q}'(t_n)$ is identical in form, but with components of $f'\{t\}$ replacing components of $f\{t\}$ and

$$[\mathcal{T}_k]^{\beta_{k+1}}_{\alpha_k} = [U^\dagger(t_{k+1})O(t_k)U(t_k)]^{\beta_{k+1}}_{\alpha_k}. \quad (3.25)$$

3.2 Discretised Influence Functional

To calculate properties of open quantum systems the general form required for the functional $f[t]$ in Eq. (3.1) is

$$f[t] = \prod_j O_j(t_j)\mathcal{I}[t], \quad (3.26)$$

for some superoperators $O_j(t_j)$ and IF $\mathcal{I}[t]$. Then multitime observables are calculated by applying $\mathcal{Q}(t)$ to an initial density matrix and taking the trace

$$\left\langle \prod_j O_j(t_j) \right\rangle = \mathrm{Tr}[\mathcal{Q}(t)\rho(0)]. \quad (3.27)$$

A specific element of the density matrix at time t could be calculated by making all but one $O_j(t_j)$ identity matrices. Alternatively the full density matrix, $\rho(t)$, could be found by making all $O_j(t_j)$ identity matrices, in which case $\mathcal{Q}(t)$ is the propagator for the system,

$$\rho(t) = \mathcal{Q}(t)\rho(0). \quad (3.28)$$

Upon discretising time it was shown in the previous section that the elements of $\mathcal{Q}(t_n)$ can be written

$$[\mathcal{Q}(t_n)]^{\alpha_n}_{\beta_1} = \prod_{k=1}^{n-1} [\mathcal{T}_k]^{\beta_{k+1}}_{\alpha_k} [\mathcal{I}\{t\}]^{\alpha_n\alpha_{n-1}\ldots\alpha_1}_{\beta_n\beta_{n-1}\ldots\beta_1} =$$

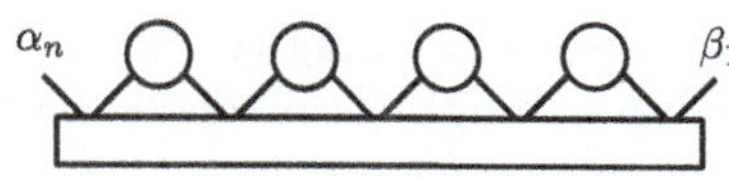

, (3.29)

where the elements of each 2-leg tensor $\mathcal{T}_k$ are as given in Eq. (3.18) and $\mathcal{I}\{t\}$ is the discretised IF, an operator in $\mathcal{G}_\Delta$ represented as a $2n$-leg tensor. The TN representation of $\mathcal{I}\{t\}$ has the form Eq. (3.12) and is in fact a process tensor [10] for the open system and for the special case of a system coupled to a HB through an Abelian interaction this has been shown explicitly [11]. More generally, $\mathcal{I}\{t\}$ is an instance of a quantum comb [12]. It is also interesting that the construction of $\mathcal{I}\{t\}$ as an operator in the product space $\mathcal{G}_\Delta$ bears semblance to the concept of superdensity operators [13]. Investigation into such connections will be the subject of future work.

The purpose of this section is to show how the single $2n$-leg tensor representing the discretised IF can be decomposed in terms of a number of much smaller tensors. Recall that any IF can be written as an exponential, Eq. (2.30), whose argument is the integral over a time-nonlocal Liouvillian functional, $\mathcal{L}_A[t]$, that has the general form Eq. (2.35). This Liouvillian is defined by a power series in the interaction potential, V, and it is assumed now that this power series has been truncated to Nth order such that $\mathcal{L}_A[t]$ is known in closed form. It is then possible to write

$$\int_0^t \mathcal{L}_A[\tau_1]d\tau_1 = \int_0^t \int_0^{\tau_1} \cdots \int_0^{\tau_{N-1}} d\tau_1 \ldots d\tau_N g(\tau_1, \ldots, \tau_N), \tag{3.30}$$

for some function of system operators $g(\tau_1, \ldots, \tau_N)$. Truncating the Liouvillian to second order, $N = 2$, corresponds to a Gaussian approximation, which is exact for Gaussian environments. Using the form for $\mathcal{L}_A[t]$ given by Eqs. (2.35) and (2.36) and truncating to $N = 2$ gives

$$g(\tau_1, \tau_2) = -ix^-(\tau_1)\langle F(\tau_1)\rangle\delta(\tau_1 - \tau_2) \\ - x^-(\tau_1)\Re[C(\tau_1, \tau_2)]x^-(\tau_2) - ix^-(\tau_1)\Im[C(\tau_1, \tau_2)]x^+(\tau_2). \tag{3.31}$$

Without yet making an approximation, the integral in Eq. (3.30) can be expressed as a sum

$$\int_0^t \int_0^{\tau_1} \cdots \int_0^{\tau_{N-1}} d\tau_1 \ldots d\tau_N g(\tau_1, \ldots, \tau_N) = \sum_{k_1=1}^{n} \sum_{k_2=1}^{k_1} \cdots \sum_{k_N=1}^{k_{N-1}} g_{k_1\ldots k_N}, \tag{3.32}$$

with

$$g_{k_1\ldots k_N} = \int_{t_{k_1-1}}^{t_{k_1}} \int_{t_{k_2-1}}^{t_{k_2}} \cdots \int_{t_{k_N-1}}^{t_{k_N}} d\tau_1 \ldots d\tau_N g(\tau_1, \ldots, \tau_N). \tag{3.33}$$

Then $g_{k_1\ldots k_N}$ is a function of N time-local operators $x(\tau_l)$ whose time variables take values in the intervals $\tau_l \in [t_{k_l-1}, t_{k_l}]$ for $1 < l < N$. The discretisation approximation is made by setting $x(\tau_l) \approx x(t_{k_l})$, such that each $g_{k_1 \ldots k_N}$ can be represented as an operator acting in the product space $\mathcal{G}(t_{k_1}) \otimes \mathcal{G}(t_{k_2}) \otimes \cdots \otimes \mathcal{G}(t_{k_N}) \in \mathcal{G}_\Delta$. For $N = 2$ this approximation reads

$$g_{k_1 k_2} \approx \begin{cases} -ix^-(t_{k_1})\langle F(t_{k_1})\rangle\Delta - x^-(t_{k_1})(\Re[\eta_{k_1 k_2}]x^-(t_{k_2}) + i\Im[\eta_{k_1 k_2}]x^+(t_{k_2})) & k_1 = k_2 \\ -x^-(t_{k_1})(\Re[\eta_{k_1 k_2}]x^-(t_{k_2}) + i\Im[\eta_{k_1 k_2}]x^+(t_{k_2})) & k_1 \neq k_2 \end{cases} \tag{3.34}$$

with

$$\eta_{k_1 k_2} = \begin{cases} \int_{t_{k_1 - 1}}^{t_{k_1}} \int_{t_{k_1 - 1}}^{\tau_1} C(\tau_1, \tau_2) d\tau_2 d\tau_1 & k_1 = k_2 \\ \int_{t_{k_1 - 1}}^{t_{k_1}} \int_{t_{k_2 - 1}}^{t_{k_2}} C(\tau_1, \tau_2) d\tau_2 d\tau_1 & k_1 \neq k_2 \end{cases}. \tag{3.35}$$

Using Eq. (3.30) to represent the IF in terms of $g(\tau_1, \ldots, \tau_N)$ and making the discretisation approximation allows one to write

$$\mathcal{I}[t_n] \approx \mathcal{I}\{t\} = \prod_{k_1=1}^{n} \prod_{k_2=1}^{k_1} \cdots \prod_{k_N=1}^{k_{N-1}} \mathcal{A}_{k_1 \ldots k_N}, \tag{3.36}$$

where the exponential of the sums in Eq. (3.30) has been written as a product of the operators

$$\mathcal{A}_{k_1 \ldots k_N} = e^{g_{k_1 \ldots k_N}}, \tag{3.37}$$

which act in the subspace $\mathcal{G}(t_{k_1}) \otimes \mathcal{G}(t_{k_2}) \otimes \cdots \otimes \mathcal{G}(t_{k_N}) \in \mathcal{G}_\Delta$. Each operator $\mathcal{A}_{k_1 \ldots k_N}$ can be decomposed onto the basis operators of its respective subspace

$$\mathcal{A}_{k_1 \ldots k_N} = \hat{e}_{\alpha_{k_1}}^{\beta_{k_1}} \otimes \cdots \otimes \hat{e}_{\alpha_{k_N}}^{\beta_{k_N}} [\mathcal{A}_{k_1 \ldots k_N}]_{\beta_{k_1} \ldots \beta_{k_N}}^{\alpha_{k_1} \ldots \alpha_{k_N}}, \tag{3.38}$$

and can be represented as a, at most, $2N$-leg tensor. Multiplication between different $\mathcal{A}_{k_1 \ldots k_N}$ is then standard matrix multiplication in $\mathcal{G}_\Delta$ and can be represented by tensor contractions. For example, in the case $N = 2$ the operator $\mathcal{A}_{k_1 k_2}$ with $k_1 \neq k_2$ acts simultaneously in the spaces $\mathcal{G}(t_{k_1})$ and $\mathcal{G}(t_{k_2})$, has elements $[\mathcal{A}_{k_1 k_2}]_{\beta_{k_1} \beta_{k_2}}^{\alpha_{k_1} \alpha_{k_2}}$, and is represented diagrammatically like

$$[\mathcal{A}_{k_1 k_2}]_{\beta_{k_1} \beta_{k_2}}^{\alpha_{k_1} \alpha_{k_2}} =$$

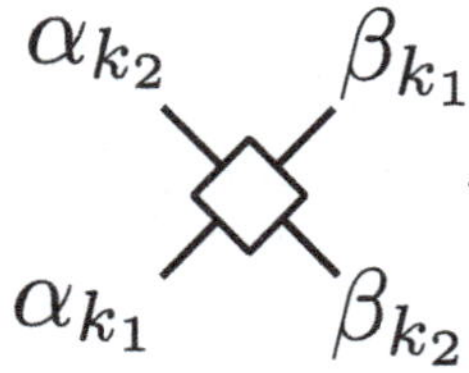

. (3.39)

The operator $\mathcal{A}_{k_1 k_1}$, however, acts only in a single subspace $\mathcal{G}(t_{k_1})$ so is a 2-leg tensor with elements $[\mathcal{A}_{k_1 k_1}]_{\beta_{k_1}}^{\alpha_{k_1}}$ and diagram

$$[\mathcal{A}_{k_1 k_1}]^{\alpha_{k_1}}_{\beta_{k_1}} = \tag{3.40}$$

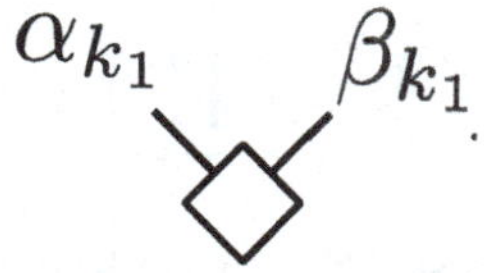

Multiplication between $\mathcal{A}_{k_1 k_1}$ and $\mathcal{A}_{k_1 k_2}$ only occurs in subspaces they both act in, in this case $\mathcal{G}(t_{k_1})$, and diagrammatically there is a connecting leg between them for each subspace they both act in. Thus,

$$\mathcal{A}_{k_1 k_1} \mathcal{A}_{k_1 k_2} = \tag{3.41}$$

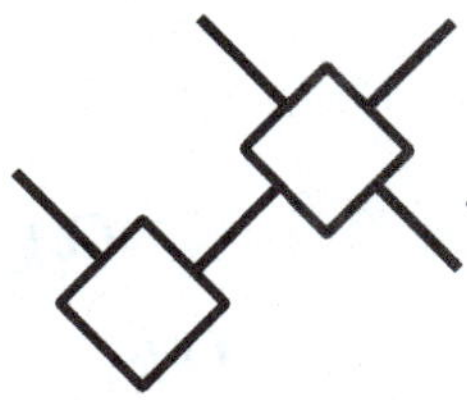

The representation of a discretised IF in terms of products of $\mathcal{A}_{k_1 \ldots k_N}$ operators thus allows it to decomposed into a TN of many smaller tensors. To see how this works first consider the case $N = 1$, making the IF time-local. In this case each $\mathcal{A}_{k_1}$ acts in a independent subspace of $\mathcal{G}_\Delta$. That is, as an operator in $\mathcal{G}_\Delta$, $\mathcal{I}\{x(t)\}$ is simply a tensor product of the $\mathcal{A}_{k_1}$,

$$\mathcal{I}\{t\} = \mathcal{A}_n \otimes \mathcal{A}_{n-1} \otimes \cdots \otimes \mathcal{A}_1, \tag{3.42}$$

and so has a TN representation as n disconnected 2-leg tensors

$$[\mathcal{I}\{t\}]^{\alpha_n \ldots \alpha_k \ldots \alpha_1}_{\beta_n \ldots \beta_k \ldots \beta_1} = [\mathcal{A}_n]^{\alpha_n}_{\beta_n} \ldots [\mathcal{A}_k]^{\alpha_k}_{\beta_k} \ldots [\mathcal{A}_1]^{\alpha_1}_{\beta_1} = \tag{3.43}$$

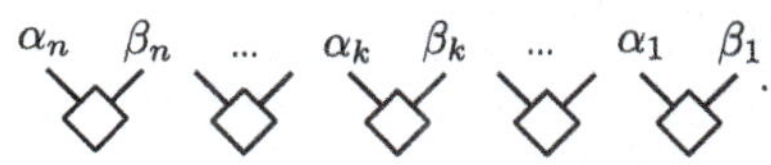

Then the operator $\mathcal{Q}(t_n)$ has the TN representation

$$[\mathcal{Q}(t_n)]^{\alpha_n}_{\beta_1} = \tag{3.44}$$

where the circular 2-leg tensors are $\mathcal{T}_k$.

Now consider an IF with $N = 2$ such that

$$\mathcal{I}\{t\} = \prod_{k_1=1}^{n} \prod_{k_2=1}^{k_1} \mathcal{A}_{k_1 k_2}, \tag{3.45}$$

and each $\mathcal{A}_{k_1 k_2}$ acts in two subspaces of $\mathcal{G}_\Delta$. For a given value of k_1 the object,

$$\tilde{\mathcal{A}}_{k_1 \ldots 1} = \prod_{k_2=1}^{k_1} \mathcal{A}_{k_1 k_2} = \mathcal{A}_{k_1 k_1} \ldots \mathcal{A}_{k_1 1}, \tag{3.46}$$

is a $2k_1$-leg tensor whose TN diagram found by contracting $k_1 - 1$ 4-leg tensors and one 2-leg tensor,

$$[\tilde{\mathcal{A}}_{k_1 \ldots 1}]^{\alpha_{k_1} \ldots \alpha_1}_{\beta_{k_1} \ldots \beta_1} =$$

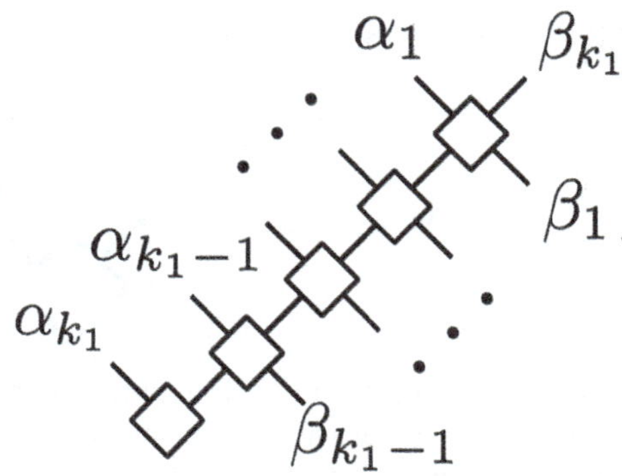

(3.47)

The IF in Eq. (3.45) is then written

$$\mathcal{I}\{t\} = \prod_{k_1=1}^{n} \tilde{\mathcal{A}}_{k_1 \ldots 1}, \tag{3.48}$$

which is represented by contractions between n tensors of the form Eq. (3.47), resulting in

$$[\mathcal{I}\{t\}]^{\alpha_n \ldots \alpha_k \ldots \alpha_1}_{\beta_n \ldots \beta_k \ldots \beta_1} =$$

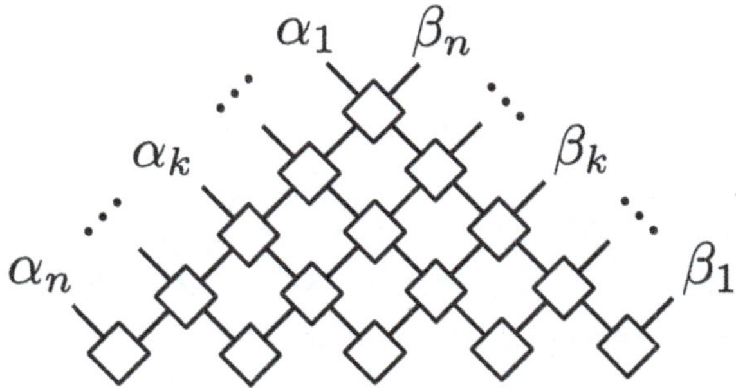

(3.49)

Then the elements of $\mathcal{Q}(t_n)$ are

$$[\mathcal{Q}(t_n)]^{\alpha_n}_{\beta_1} =$$

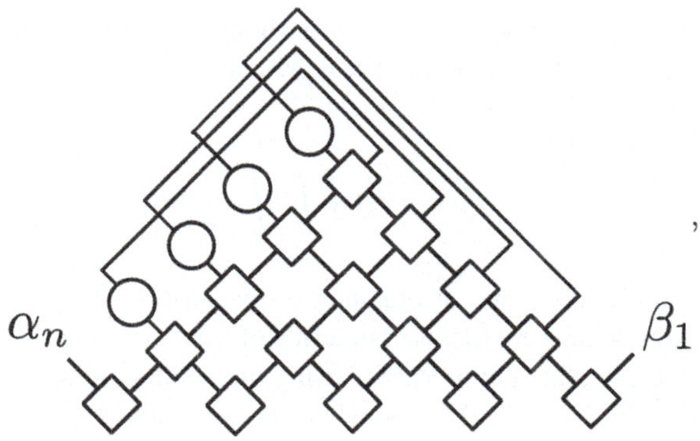

, (3.50)

where again the circular 2-leg tensors are $\mathcal{T}_k$.

The TN representation of IFs for $N > 2$ shall not be given explicitly. It is worth pointing out, however, that any IF with $N > 2$ can be approximated by one with $N = 2$ and so can approximately represented by a TN of the form Eq. (3.49). This is achieved by using a more general form of the Redfield approximation that was discussed in Sect. 2.1.2.2. Using the current notation the Redfield approximation as given by Eq. (2.63) would read

$$\int_0^t \int_0^{\tau_1} \cdots \int_0^{\tau_{N-1}} d\tau_1 d\tau_2 \ldots d\tau_N g(\tau_1, \tau_2, \tau_3 \ldots, \tau_N) \approx \int_0^t d\tau_1 \tilde{g}(\tau_1), \quad (3.51)$$

with

$$\tilde{g}(\tau_1) = \int_0^{\tau_1} \cdots \int_0^{\tau_{N-1}} d\tau_2 \ldots d\tau_N g(\tau_1, \tau_2, \tau_3 \ldots, \tau_N). \quad (3.52)$$

The approximation being made here is to perform the integrals over τ_2 to τ_N before the time ordering operator is applied leaving just a time-local object, $\tilde{g}(\tau_1)$, equivalent to a time-local Liouvillian. If instead the approximation is made by performing the integrals only over τ_3 to τ_N then the result is,

$$\int_0^t \int_0^{\tau_1} \cdots \int_0^{\tau_{N-1}} d\tau_1 d\tau_2 \ldots d\tau_N g(\tau_1, \tau_2, \tau_3 \ldots, \tau_N) \approx \int_0^t \int_0^{\tau_1} d\tau_2 \tilde{g}(\tau_1, \tau_2), \quad (3.53)$$

with

$$\tilde{g}(\tau_1, \tau_2) = \int_0^{\tau_2} \cdots \int_0^{\tau_{N-1}} d\tau_3 \ldots d\tau_N g(\tau_1, \tau_2, \tau_3 \ldots, \tau_N). \quad (3.54)$$

This approximation can be made for arbitrary N and so can be used to nonperturbatively put arbitrary IFs in the form of those with $N = 2$, allowing their representation by a TN of the form Eq. (3.49). The effectiveness of this approximation Eq. (3.53) is yet to be investigated in practice, though, like the Redfield approximation, it is expected to be justifiable when temporal correlations amongst system operators are weak, i.e. when $[x(t), x(t')] \approx 0$.

Before concluding this section it should be stressed that the procedure carried out to arrive at the triangular TN representation of $\mathcal{I}[t]$, Eq. (3.49), is valid for any functional of the form

$$f[t] = \exp\left(\int_0^t \int_0^{\tau_1} d\tau_2 d\tau_1 g(\tau_1, \tau_2)\right). \tag{3.55}$$

For example, elements of the state propagator e^{-iHt} of the collective system and environment can be expressed in terms of time ordering of a functional of this form, so can be calculated numerically using the results of this chapter. More specifically, assuming $H = H_A + H_B + V$ where H_A, H_B, and V, are the system, environment, and interaction Hamiltonians, elements of the state propagator can be written

$$\langle a| \langle b| e^{-iHt} |b'\rangle |a'\rangle = \langle a| \overleftarrow{\mathrm{T}}_A \left(e^{-i\int_0^t H_A(t')dt'} f[t]\right) |a'\rangle, \tag{3.56}$$

where $H_A(t) = H_A$, $|a\rangle$ and $|a'\rangle$ ($|b\rangle$ and $|b'\rangle$) are states vectors of the system (bath), and time ordering is of operators, not superoperators, of the system. Taking $\langle b| V |b'\rangle = 0$ the functional $f[t]$ is an exponential with argument

$$\log(f[t]) = -\int_0^t \int_0^{\tau_1} d\tau_1 d\tau_2 \langle b(t)| V(\tau_1)V(\tau_2) |b'(0)\rangle + \mathcal{O}(V^4), \tag{3.57}$$

where $|b(t)\rangle$, and $V(t)$ have been transformed into the interaction frame with respect to H_B. In the case that terms $\mathcal{O}(V^4)$ are neglected, or vanish identically as in the case that the environment is Gaussian [8], $f[t]$ has the form Eq. (3.55) with

$$g(\tau_1, \tau_2) = -\langle b(t)| V(\tau_1)V(\tau_2) |b'(0)\rangle. \tag{3.58}$$

Another example is the reduced state of a collective equilibrium state of the system and environment at inverse temperature β. Defining the collective and environment partition functions, $Z = \mathrm{Tr}e^{-\beta H}$ and $Z_B = \mathrm{Tr}_B e^{-\beta H_B}$ where Tr_B is the partial trace over the environment, the reduced equilibrium state takes the general form

$$\rho_A(\beta) = Z^{-1}\mathrm{Tr}_B[e^{-\beta H}] = Z_B Z^{-1} \overleftarrow{\mathrm{T}}_A \left(e^{-\int_0^\beta H_A(\tau)d\tau} f[\beta]\right), \tag{3.59}$$

where time ordering of operators is now along the imaginary time axis from 0 to β and $H_A(\tau) = H_A$. For an environment with $Z_B^{-1}\mathrm{Tr}[Ve^{-\beta H_B}] = 0$, the functional $f[\beta]$ is an exponential with argument

$$\log(f[\beta]) = \int_0^\beta \int_0^{\tau_1} Z_B^{-1}\mathrm{Tr}_B\left[\overleftarrow{\mathrm{T}}_B V(\tau_1)V(\tau_2)e^{-\int_0^\beta H_B(\tau')d\tau'}\right] + \mathcal{O}(V^4), \tag{3.60}$$

where $V(\tau) = V$ and $H_B(\tau) = H_B$. In the case that terms $\mathcal{O}(V^4)$ are neglected, or vanish identically as in the case that the environment is Gaussian [14], $f[t]$ has the form Eq. (3.55) with

$$g(\tau_1, \tau_2) = Z_B^{-1}\mathrm{Tr}_B\left[\overleftarrow{\mathrm{T}}_B\, V(\tau_1)V(\tau_2)e^{-\int_0^\beta H_B(\tau')d\tau'}\right]. \tag{3.61}$$

For simple test systems, numerical evaluation of Eqs. (3.56) and (3.59) using the TN techniques described in this chapter has been successful. The results of these calculations and the potential applications of this technique will be the subject of future work.

3.3 Time-Evolving Matrix Product Operators

To make the TN shown in Eq. (3.50) amenable to efficient contraction, it is rearranged by deforming the connecting legs external to the triangular structure so that the 2-leg $\mathcal{T}_k$ tensors are imposed on the bottom row of 4-leg tensors,

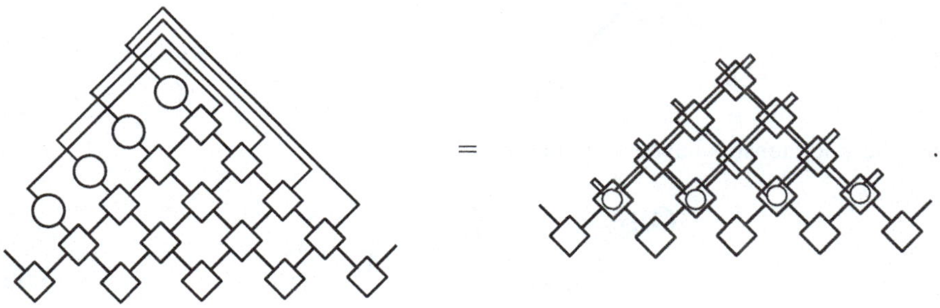

(3.62)

Now the elements of $\mathcal{Q}(t_n)$ can be written in terms of a new set of tensors,

$$[\mathcal{Q}(t_n)]^{\alpha_n}_{\beta_1} =$$

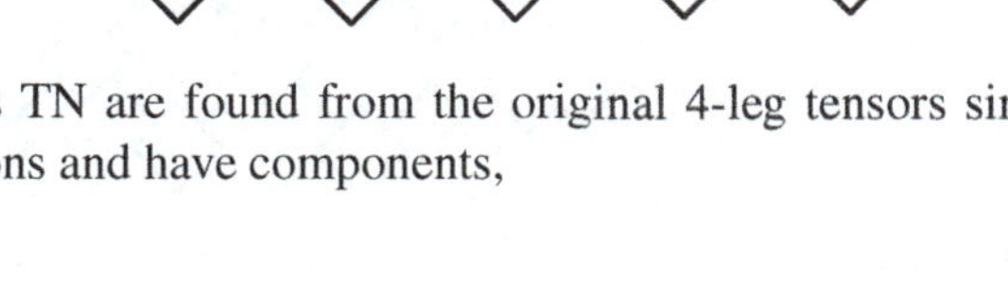

(3.63)

The 8-leg tensors in this TN are found from the original 4-leg tensors simply by inclusion of delta functions and have components,

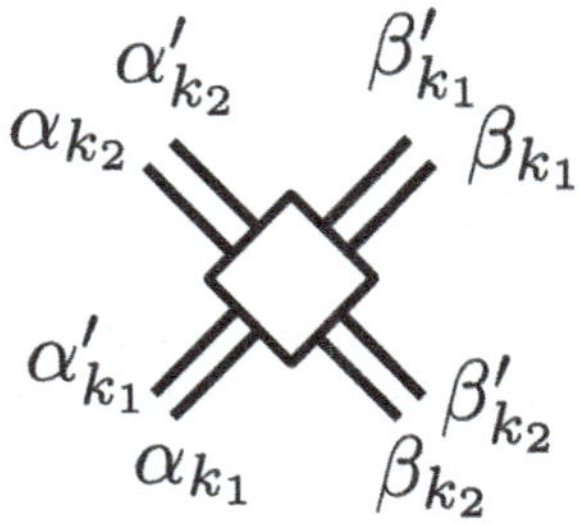

$$= \quad [\mathcal{A}_{k_1k_2}]^{\alpha_{k_1}\alpha_{k_2}}_{\beta_{k_1}\beta_{k_2}}\delta^{\alpha'_{k_1}}_{\beta'_{k_1}}\delta^{\alpha'_{k_2}}_{\beta'_{k_2}}, \tag{3.64}$$

from which the 6-leg tensors on the edges and the 4-leg tensor at the tip of triangle can be found by contracting together neighbouring legs. The 6-leg tensors on the second to bottom row of the TN have components,

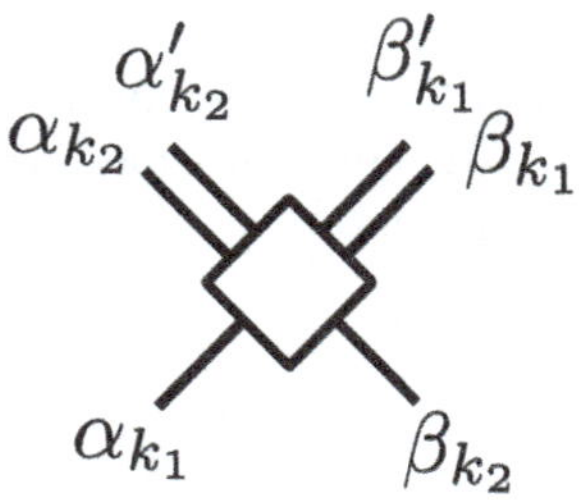

$$= \quad [\mathcal{A}_{k_1k_2}]^{\alpha_{k_1}\alpha_{k_2}}_{\beta_{k_1}\beta_{k_2}}[\mathcal{T}_{k_1}]^{\alpha'_{k_2}}_{\beta'_{k_1}}, \tag{3.65}$$

and the 2-leg tensors have components,

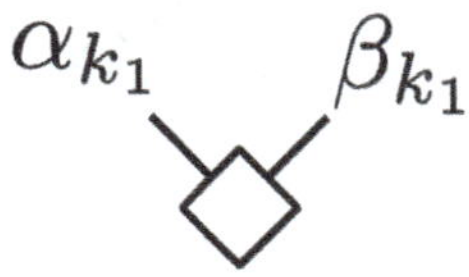

$$= [\mathcal{A}_{k_1k_1}]^{\alpha_{k_1}}_{\beta_{k_1}}, \tag{3.66}$$

As mentioned at the beginning of the previous section, the operator $\mathcal{Q}(t_n)$ is typically acted upon some density matrix in order to calculate quantities of interest. Thus, the TN whose contraction should be made efficient is

$$\mathcal{Q}(t_n)\rho(0) =$$

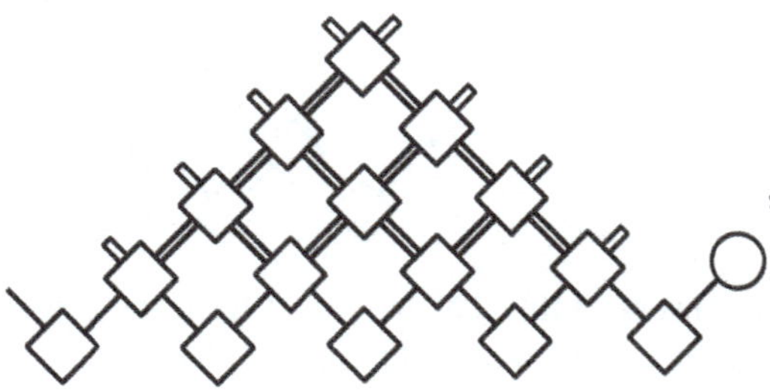

$$, \tag{3.67}$$

where the circular 1-leg tensor represents $\rho(0)$, which has been mapped to a d-dimensional vector in $\mathcal{G}$. Note that, indeed, multiplication of the operator $\mathcal{Q}(t_n)$ into the vector $\rho(0)$ has produced another vector, since the overall TN has only one free leg. This TN can now be contracted efficiently using the techniques for contracting matrix products outlined in Sect. 2.2. Performing contractions from right to left, the first one is between a 2-leg tensor and the 1-leg density matrix to produce another 1-leg tensor,

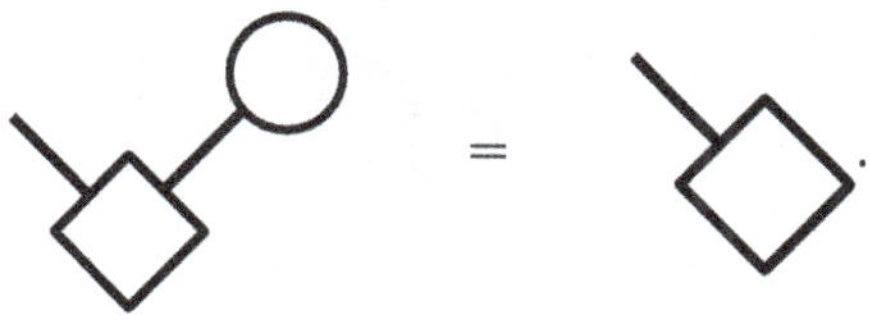

(3.68)

The next contraction is

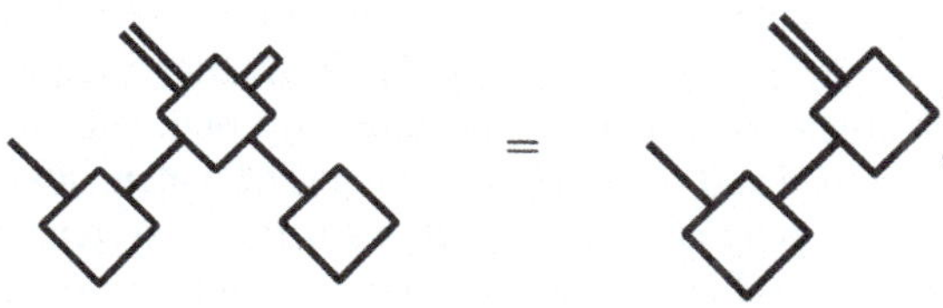

(3.69)

which has resulted in an MPS with two sites. Each contraction after this is in the form of an MPO being contracted with an MPS, which grows by one site after each contraction, and so SVDs can be performed on the MPS as contractions are ongoing. The contraction that follows Eq. (3.69) is

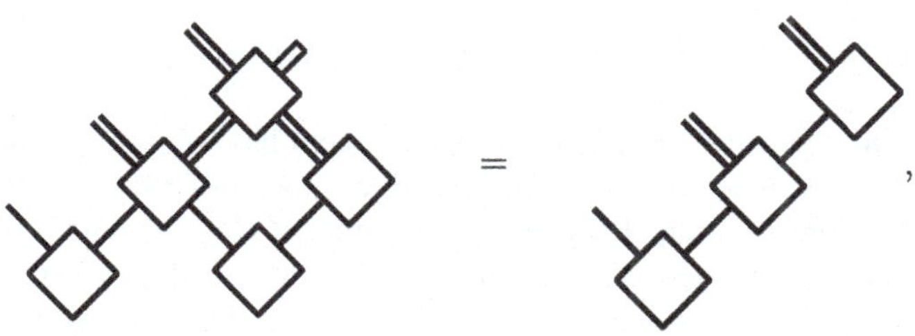

(3.70)

and so on. Notice that this process can be terminated after any number of contractions. That is, the object $\mathcal{Q}(t_k)\rho(0)$ with $k < n$ can be found instead of $\mathcal{Q}(t_n)\rho(0)$ once the MPS has been grown to have k sites simply by connecting the pairs of free legs on each site of the MPS that has a pair of legs. For example, $\mathcal{Q}(t_4)\rho(0)$ could be found from the MPS once it has been grown to four sites like,

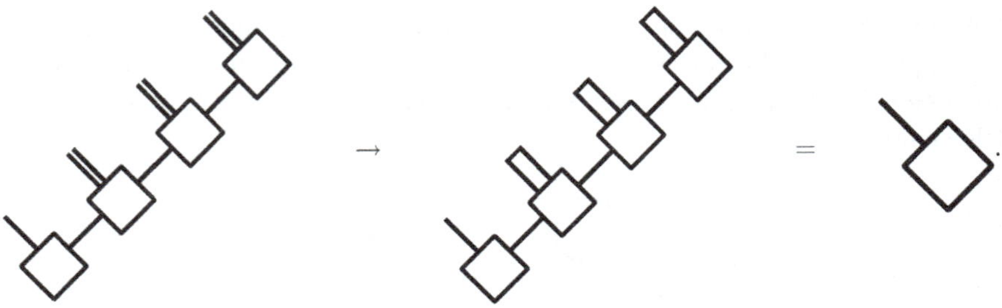

(3.71)

The explicit form of the IF for an arbitrary open system can be found using the results in Sect. 2.1.1.2 and the generalised Redfield approximation, Eq. (3.53), can be used on the IF to enable the calculation of arbitrary observables by contracting the TN representation of $\mathcal{Q}(t)\rho(0)$ given by Eq. (3.67). In principle this contraction can be carried out efficiently using the procedure just outlined, but the tensors that make up this TN are each still rather large, having $\mathcal{O}(d^8)$ elements. This size can be significantly reduced if only abelian interactions between system and environment are considered and observables are calculated in the lab frame rather than the interaction frame which has been used throughout. As discussed at the end of Sect. 3.1, the discretised IF can be written in terms of lab frame operators, $x^{\pm}(t) = x^{\pm}$, by projecting onto a time dependent basis, $\hat{e}^{\beta}_{\alpha}(t_k) = U(t_k)\hat{e}^{\beta}_{\alpha}U^{\dagger}(t_k)$, where $U(t)$ is defined as in Eq. (3.20) with $h(t) = H_A$, the Hamiltonian for the open system in question. The transformation to the lab frame from the interaction frame is then

$$\mathcal{Q}(t_n)\rho(0) \to U^{\dagger}(t_n)\mathcal{Q}(t_n)\rho(0) = \mathcal{Q}'(t_n)U^{\dagger}(t_1)\rho(0) = \mathcal{Q}'(t_n)\rho(\Delta), \tag{3.72}$$

where the initial density matrix for the open system has been propagated forward in time by $t_1 = \Delta$ under $U^{\dagger}(t)$. As discussed at the end of Sect. 3.1, the operator $\mathcal{Q}'(t)$ has TN representation identical in form to that of $\mathcal{Q}(t)$ but with different 2-leg tensors $\mathcal{T}_k$. Then $\mathcal{Q}'(t)\rho(\Delta)$ is represented by a TN identical in structure to that of $\mathcal{Q}(t)\rho(0)$ in Eq. (3.67), but with the elements of $\mathcal{T}_k$ appearing in the 6-leg tensors, Eq. (3.65), being given by Eq. (3.25) and the circular 1-leg tensor representing $\rho(\Delta)$ not $\rho(0)$.

For an abelian interaction potential in the lab frame not only are all operators time independent, $x^{\pm}(t) = x^{\pm}$, but they can be simultaneously diagonalised. This means basis operators for $\mathcal{G}$ can be chosen such that

$$[x^{\pm}]^{\alpha}_{\beta} = \delta^{\alpha}_{\beta} x^{\pm}_{\beta}, \tag{3.73}$$

where $x^{\pm}_{\beta}$ are the d eigenvalues of $x^{\pm}$. The operators $g_{k_1k_2}$ in $\mathcal{G}_{\Delta}$ are then all diagonal, as are the operators $\mathcal{A}_{k_1k_2}$. The components of $\mathcal{A}_{k_1k_2}$ thus have the form,

$$[\mathcal{A}_{k_1k_2}]^{\alpha_{k_1}\alpha_{k_2}}_{\beta_{k_1}\beta_{k_2}} = \delta^{\alpha_{k_1}}_{\beta_{k_1}}\delta^{\alpha_{k_2}}_{\beta_{k_2}} I_{\beta_{k_1}\beta_{k_2}}, \tag{3.74}$$

where $I_{\beta_{k_1}\beta_{k_2}}$ are the d^2 eigenvalues of $\mathcal{A}_{k_1 k_2}$. Inserting Eq. (3.74) into Eq. (3.64) and one finds that every leg of every 8-leg tensor in the TN, Eq. (3.63) is composed of delta functions. It is straightforward to verify that this abundance of delta functions allows one of the two contractions that connect each pair of neighbouring tensors to be performed analytically. This allows the TN in Eq. (3.63) to be simplified like

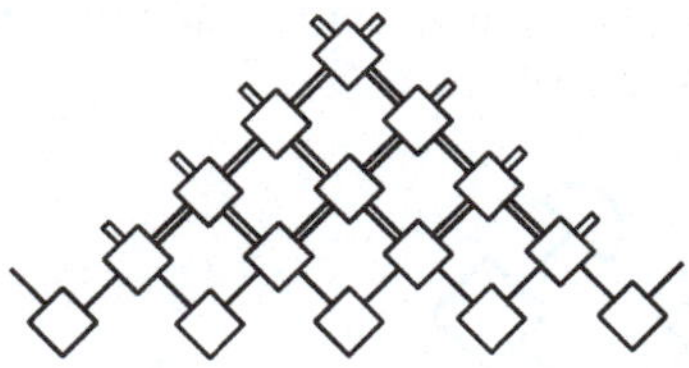
=
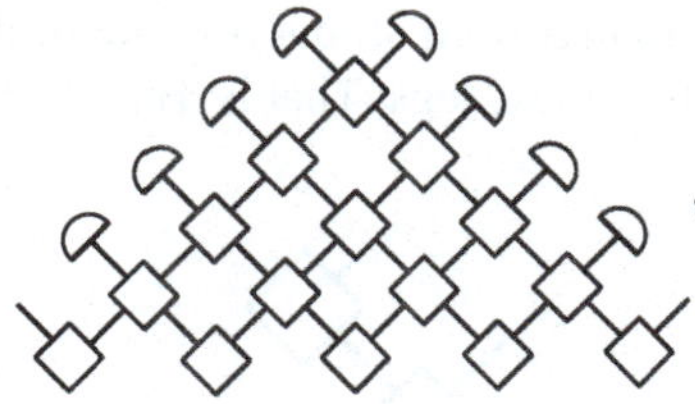
, (3.75)

where

$$= [1]_\alpha \tag{3.76}$$

is the vector whose elements are all equal to 1, the 4-leg tensors along the bottom row have elements

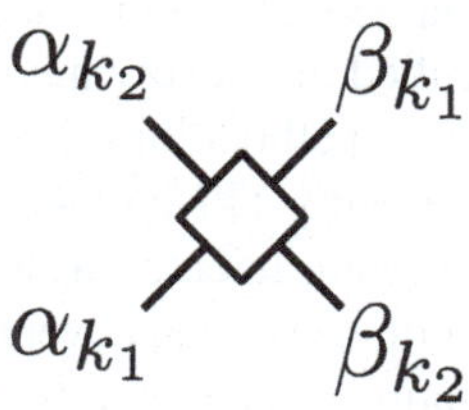

$$= \delta^{\alpha_{k_1}}_{\beta_{k_1}} \delta^{\alpha_{k_2}}_{\beta_{k_2}} I_{\beta_{k_1}\beta_{k_2}} [\mathcal{T}_{k_1}]^{\alpha_{k_2}}_{\beta_{k_1}}, \tag{3.77}$$

and for all other 4-leg tensors,

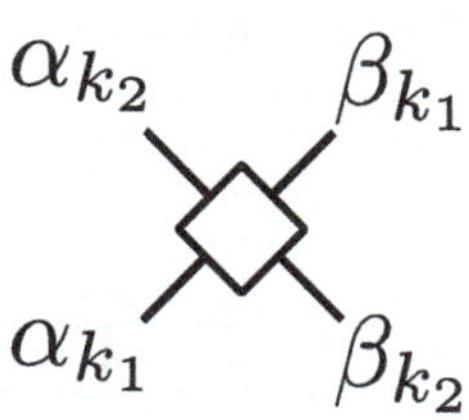

$$= \delta^{\alpha_{k_1}}_{\beta_{k_1}} \delta^{\alpha_{k_2}}_{\beta_{k_2}} I_{\beta_{k_1}\beta_{k_2}}. \tag{3.78}$$

The 2-leg tensors are given by

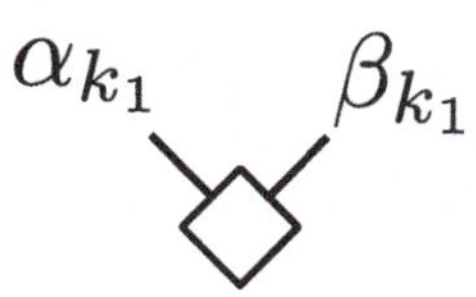

$$= I_{\beta_{k_1}\beta_{k_1}} \delta^{\alpha_{k_1}}_{\beta_{k_1}}. \tag{3.79}$$

The simplified TN on right hand side of Eq. (3.75) has the same structure as a series of MPOs as the TN on the left hand side, but the size of the individual tensors is now $\mathcal{O}(d^4)$. The procedure for contracting the TN to find the object $\mathcal{Q}'(t_n)\rho(\Delta)$ by growing an MPS is the same form as in Eqs.(3.68)–(3.70), but now using an MPO at each stage with smaller tensors. The termination of the procedure is also similar, though the object $\mathcal{Q}'(t_n)\rho(\Delta)$ is now by found connecting the single free leg protruding from all but one site of the MPS with $[1]_\alpha$, equivalent to summing over each of these legs. That is, Eq. (3.71) would now be written

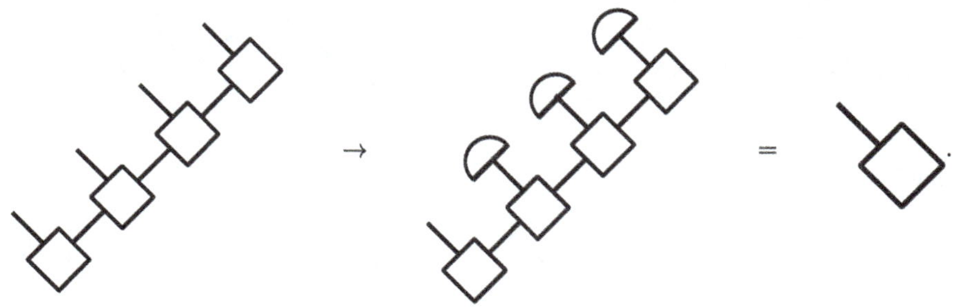

(3.80)

In the case that $\mathcal{Q}'(t_n)$ is the lab frame propagator for an open system coupled to a HB with zero mean, $\langle F(t)\rangle = 0$, through a single coupling operator, the simplified TN on right hand side of Eq. (3.75) was derived and numerically contracted in Ref. [15]. The MPOs that make up the TN in this case are called time-evolving matrix product operators (TEMPOs). From here on the collective methodology presented in this chapter will simply be referred to as TEMPO. Originally TEMPO was derived from the discretised path integral representation of a system interacting with a HB, not through the realisation of the TN representation of time ordering covered in Sect. 3.1. To see how the discretised path integral representation can be recovered note that the elements of the 4-leg tensors Eq. (3.78) that make up the bulk of the right hand side of Eq. (3.75) are still composed mostly of delta functions. This allows for further analytic contraction of the TN, at the expense of obscuring the decomposition in terms of small tensors,

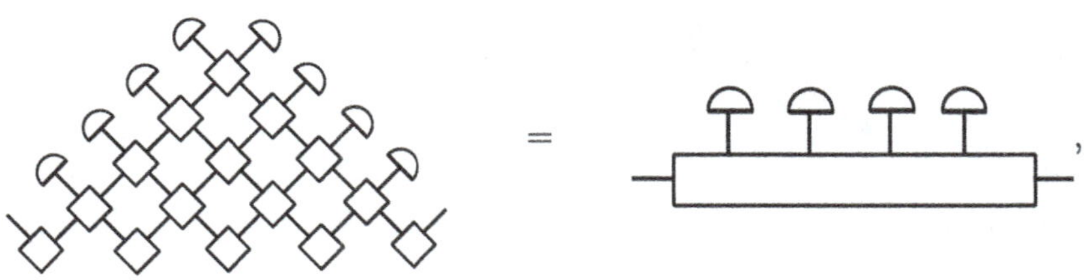

(3.81)

where

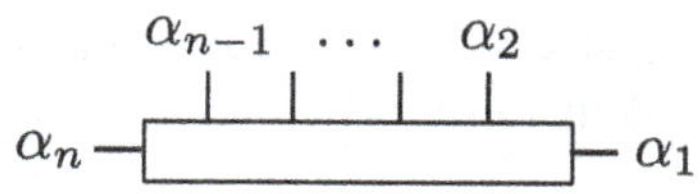

$$= \left(\prod_{k_1=1}^{n-1} [\mathcal{T}_{k_1}]_{\alpha_{k_1+1}\alpha_{k_1}} \right) \prod_{k_1=1}^{n} \prod_{k_2=1}^{k_1} I_{\alpha_{k_1}\alpha_{k_2}}. \tag{3.82}$$

with

$$[\mathcal{T}_{k_1}]_{\alpha_{k_1+1}\alpha_{k_1}} = [e^{-iH_A^- \Delta}]_{\alpha_{k_1}}^{\alpha_{k_1+1}}. \tag{3.83}$$

In the special case of a system interacting with a HB with $\langle F(t) \rangle = 0$,

$$I_{\alpha_{k_1}\alpha_{k_2}} = \exp\left(-x_{\alpha_{k_1}}^- \Re[\eta_{k_1k_2}] x_{\alpha_{k_2}}^- - i x_{\alpha_{k_1}}^- \Im[\eta_{k_1k_2}] x_{\alpha_{k_2}}^+ \right), \tag{3.84}$$

with $\eta_{k_1k_2}$ as defined in Eq. (3.35). Then the object

$$I(\alpha_n, \alpha_{n-1}, \dots, \alpha_1) = \left(\prod_{k_1=1}^{n-1} [\mathcal{T}_{k_1}]_{\alpha_{k_1+1}\alpha_{k_1}} \right) \prod_{k_1=1}^{n} \prod_{k_2=1}^{k_1} I_{\alpha_{k_1}\alpha_{k_2}}, \tag{3.85}$$

is precisely the discretised path integral representation of the Feynman-Vernon IF [3, 16], and the RDM at time t_n is found through

$$[\rho(t_n)]_{\alpha_n} = \sum_{\alpha_{n-1},\dots,\alpha_1} I(\alpha_n, \alpha_{n-1}, \dots, \alpha_1)[\rho(\Delta)]_{\alpha_1}. \tag{3.86}$$

3.4 Finite Memory Approximation

The previous section has shown that it is possible to efficiently evaluate time ordered IFs by contracting a TN. Efficiency of the technique can be further improved by making a finite memory approximation to reduce the number of tensors that make up such TNs. Consider that an IF has been put into the $N = 2$ form by making the Gaussian approximation. In this case the range of time-nonlocality in the IF is determined by the response function, $C(t, t')$, of the environment of the open system. As discussed in Sect. 2.1.1.2, for any well behaved environment this response function should vanish as $t - t'$ becomes large. Thus, the operators $g_{k_1k_2}$, as defined in Eq. (3.34), should also vanish for large $k_1 - k_2$. In turn this implies that the 4-leg tensors, $\mathcal{A}_{k_1k_2}$, constructed from $g_{k_1k_2}$ via Eq. (3.37), become identity operators in $\mathcal{G}_\Delta$. The finite memory approximation is then

$$[\mathcal{A}_{k_1k_2}]_{\beta_{k_1}\beta_{k_2}}^{\alpha_{k_1}\alpha_{k_2}} \approx \delta_{\beta_{k_1}}^{\alpha_{k_1}} \delta_{\beta_{k_2}}^{\alpha_{k_2}}, \tag{3.87}$$

when $k_1 - k_2 > K$ for suitably large K. This is equivalent to making a hard cutoff in the response function, $C(t, t') = 0$ for $t - t' > K\Delta$. Ideally K should be chosen such that $K\Delta \geq \tau_c$ where τ_c is the relaxation timescale of the environment. The approximation given by Eq. (3.87) is to decompose the 4-leg tensor $\mathcal{A}_{k_1k_2}$ into two independent 2-leg tensors whose components are delta functions,

, (3.88)

thus making the analytic contraction of such tensors trivial. Making the finite memory approximation then allows partial contraction of the TN given in Eq. (3.63). If $K = 2$, for example, this would result in the simplification,

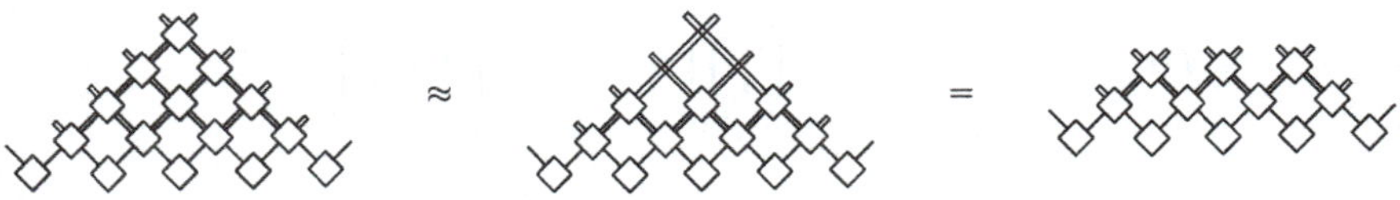

(3.89)

and similarly in the case of an abelian interaction

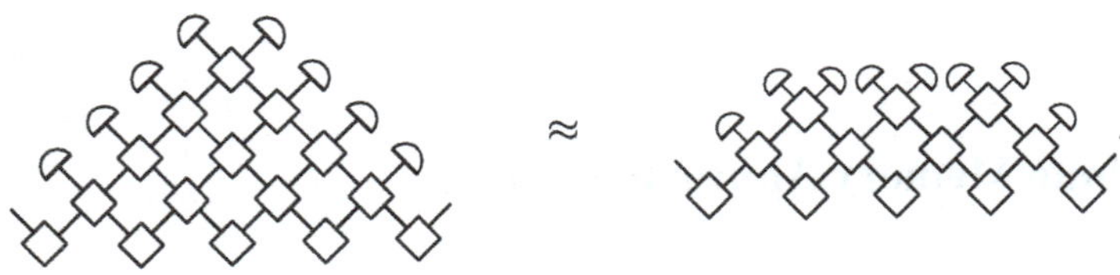

. (3.90)

The procedure for contracting such TNs now consists of a series of initial contractions, during which an MPS is grown in length until it consists of K sites, followed by successive contractions with MPOs with $K + 1$ sites. In the case of an abelian interaction and $K = 4$, for example, the MPS would initially be grown to be 4 sites long and then the rest of the contractions would all take the form

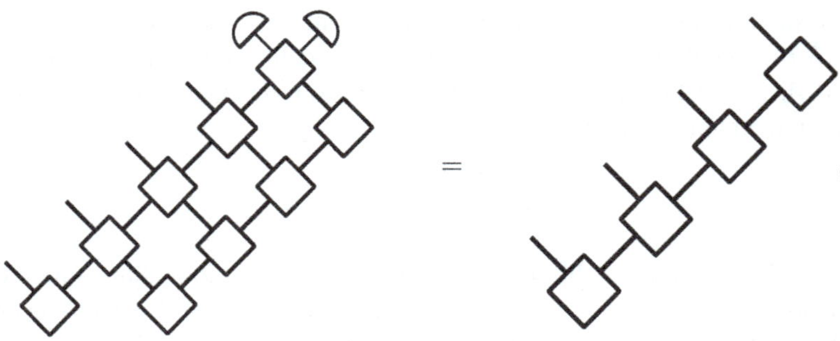

, (3.91)

meaning the MPS is never composed of more than 4 tensors.

The finite memory approximation was originally introduced as a means of making the discretised path integral for a system coupled to a HB, Eq. (3.86), amenable to numerical evaluation, the resulting technique being called the quasi-adiabatic propagator path integral (QUAPI) [3, 16, 17]. The TN representation of QUAPI can be recovered from Eq. (3.90) by contracting the separate sites of each MPO together,

(3.92)

The contraction of this TN no longer takes the form of repeated contraction between an MPO and MPS, but between a tensor propagator and a so-called augmented density tensor (ADT). The ADT is recovered from the corresponding MPS by contracting all sites of the MPS together. In the same way the MPS is grown in TEMPO, the ADT is grown from a 1-leg tensor to a K-leg tensor, before being contracted repeatedly with $2K$-leg tensor propagators. Since the MPO/MPS representation of the problem is not realised in QUAPI, the ADT always has size d^K which cannot be reduced and limits the application of the technique to problems where $K = \mathcal{O}(10)$ can be justified. A comparison between TEMPO and QUAPI will be given in Sect. 3.6 to show how the MPS representation of the ADT makes TEMPO significantly more efficient.

3.5 Improved Finite Memory Approximation

Although the finite memory approximation is well established, the kind of error it induces has not been quantified analytically. One way to understand the consequences of making the finite memory approximation is to explore what happens for an exactly soluble model. Consider the pure dephasing TLS described by the Liouvillian Eq. (2.90). The exact evolution of the transition amplitude is given by Eq. (2.91) and is determined by the decoherence rate $\gamma_p(t) = \Re[\dot{\eta}_p(t)]$, where $\eta_p(t)$ defined by Eq. (2.45). In this case the response function is that of a HB so is stationary, $C(\tau = t - t') = C(t, t')$. Letting $K\Delta = \tau_c$ for some τ_c, the finite memory approximation can be written $C(\tau) \approx C(\tau)\theta(\tau_c - \tau)$, where $\theta(t)$ is the step function. Inserting this approximated response function into the definition for $\eta_p(t)$ results in a decoherence rate for the pure dephasing TLS given by

$$\gamma_p(t) \approx \theta(\tau_c - t)\gamma_p(t) + \theta(t - \tau_c)\gamma_p(\tau_c). \tag{3.93}$$

This says that, under the finite memory approximation, the decoherence rate behaves identically to the exact solution at times t less than the memory cutoff time τ_c, but for t greater than τ_c it assumes the constant value $\gamma_p(\tau_c)$. A comparison between the exact result for $\gamma_p(t)$ and that using the finite memory approximation, Eq. (3.93), is shown in Fig. 3.1a for Ohmic and superohmic SDs. In each case the decoherence rate calculated using the finite memory approximation assumes a constant value beyond the memory time τ_c. For the Ohmic SD this results in a larger rate than the exact value while for superohmic this results in a rate that assumes a constant negative value for $t > \tau_c$. The exact evolution of the transition amplitude of a pure dephasing TLS and that using the finite memory approximation is shown in Fig. 3.1b. For the case that the SD is Ohmic the amplitude decays away faster than the exact result, while for the superohmic SD it grows exponentially for all $t > \tau_c$, which will eventually make it unphysical. In fact this unphysicality will occur for superohmic SDs using any large finite value memory cutoff τ_c, since $\gamma_p(t)$ in this case approaches zero from below. This demonstrates that the finite memory approximation can cause rather serious error in its current form.

To see how the finite memory approximation can be improved, consider how it can alternatively be expressed [18]

$$\begin{aligned}\int_0^t \int_0^{\tau_1} g(\tau_1, \tau_2) d\tau_2 d\tau_1 \\ = \int_0^{\tau_c} \int_0^{\tau_1} g(\tau_1, \tau_2) d\tau_2 d\tau_1 + \int_{\tau_c}^t \int_{\tau_1-\tau_c}^{\tau_1} g(\tau_1, \tau_2) d\tau_2 d\tau_1 + \int_{\tau_c}^t \int_0^{\tau_1-\tau_c} g(\tau_1, \tau_2) d\tau_2 d\tau_1 \\ \approx \int_0^{\tau_c} \int_0^{\tau_1} g(\tau_1, \tau_2) d\tau_2 d\tau_1 + \int_{\tau_c}^t \int_{\tau_1-\tau_c}^{\tau_1} g(\tau_1, \tau_2) d\tau_2 d\tau_1,\end{aligned} \tag{3.94}$$

where $\tau_c = K\Delta$ and the approximately equals symbol signifies the finite memory approximation. The integral has been separated in the second line to isolate the part over the region $\tau_1 - \tau_2 > \tau_c$, the rightmost term, and in the second line this part has been discarded. The overall integral region is visualised in Fig. 3.1c. The first, second, and third terms in the second line are shown by regions A, B, and C in Fig. 3.1c.

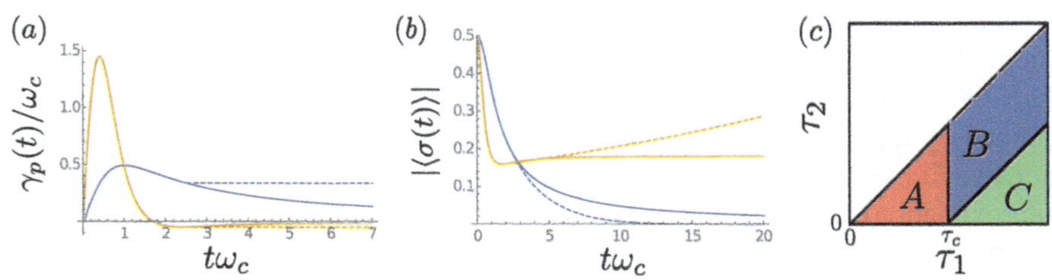

Fig. 3.1 **a** The decoherence rate of a pure dephasing TLS with $\beta^{-1} = 0$, using a SD of the form Eq. (2.44) with $\alpha = 1$ for $\nu = 1$ (blue) and $\nu = 3$ (orange). Dashed lines are the exact results, solid lines are using a finite memory approximation with $\tau_c\omega_c = 2.5$. **b** Magnitude of the transition amplitude of a pure dephasing TLS with $\beta^{-1} = 0$, using a SD of the form Eq. (2.44) with $\alpha = 1$ for $\nu = 1$ (blue) and $\nu = 3$ (orange). Dashed lines are the exact results, solid line are using a finite memory approximation with $\tau_c\omega_c = 2.5$. **c** The integration region over time dependent operators that make up an influence functional. The memory cutoff time τ_c is marked on the τ_1 axis

Recall that the exact solution for the dynamics of the RDM for pure dephasing models is found making the Redfield approximation. Instead of making the finite memory approximation by setting the last term on the first line of Eq. (3.87) equal to zero one could instead make a similar approximation to the Redfield approximation on it

$$\begin{aligned}\int_0^t \int_0^{\tau_1} g(\tau_1, \tau_2)d\tau_2 d\tau_1 \\ &= \int_0^{\tau_c} \int_0^{\tau_1} g(\tau_1, \tau_2)d\tau_2 d\tau_1 + \int_{\tau_c}^t \int_{\tau_1-\tau_c}^{\tau_1} g(\tau_1, \tau_2)d\tau_2 d\tau_1 + \int_{\tau_c}^t \int_0^{\tau_1-\tau_c} g(\tau_1, \tau_2)d\tau_2 d\tau_1 \\ &\approx \int_0^{\tau_c} \int_0^{\tau_1} g(\tau_1, \tau_2)d\tau_2 d\tau_1 + \int_{\tau_c}^t \int_{\tau_1-\tau_c}^{\tau_1} g(\tau_1, \tau_2)d\tau_2 d\tau_1 + \int_{\tau_c}^t \tilde{g}(\tau_1, \tau_1 - \tau_c)d\tau_1,\end{aligned} \tag{3.95}$$

where

$$\tilde{g}(\tau_1, \tau_1 - \tau_c) = \int_0^{\tau_1-\tau_c} g(\tau_1, \tau_2)d\tau_2. \tag{3.96}$$

Like for the Redfield approximation, Eq. (3.51), and its generalisation, Eq. (3.53), the approximation being made in the second line of Eq. (3.95) is to perform time integration before application of the time ordering operator. In the case $\tau_c = \tau_1$ no approximation has been made, while for $\tau_c = 0$ the finite memory approximation given by Eq. (3.95) is equivalent to the full Redfield approximation, Eq. (3.53). This means that the approximation being made in Eq. (3.95) has no impact on the dynamics of the RDM in pure dephasing models, since in this case the full Redfield approximation is made regardless and is exact.

Discretising the right hand side of Eq. (3.95) gives

$$\int_0^t \int_0^{\tau_1} g(\tau_1, \tau_2)\tau_2\tau_1 \approx \sum_{k_1=1}^{K} \sum_{k_2=1}^{k_1} g_{k_1k_2} + \sum_{k_1=K+1}^{K} \sum_{k_2=k_1-K}^{k_1} g_{k_1k_2} + \sum_{k_1=K+1}^{K} \tilde{g}_{k_1k_1-K}. \tag{3.97}$$

where

$$\tilde{g}_{k_1k_1-K} = \int_{t_{k_1}-1}^{t_{k_1}} \tilde{g}(\tau_1, \tau_1 - \tau_c)d\tau_1. \tag{3.98}$$

Notice that there are no terms in any of the sums that contain operators with $k_1 - k_2 > K$, such that it is structurally equivalent to having made $g_{k_1k_2} = 0$ for $k_1 - k_2 > K$, as in the standard finite memory approximation. However, for all $k_1 > K$ there are now two operators indexed by k_1 and $k_1 - K$ which are $g_{k_1k_1-K}$ and $\tilde{g}_{k_1k_1-K}$. Then making the finite memory approximation according to Eq. (3.95) is equivalent simultaneously to taking Eq. (3.87) and to setting

$$\mathcal{A}_{k_1k_1-K} = e^{g_{k_1k_1-K}+\tilde{g}_{k_1k_1-K}}. \tag{3.99}$$

Thus, by using Eq. (3.95) instead of making a hard cutoff in the environment response function, the same reductions in the size of the TN can be made but the potentially severe error that is apparent when the hard cutoff is made in pure dephasing models is completely eliminated. Furthermore, implementing this improved finite memory approximation only requires the modification of a small subset of tensors in the reduced TN one obtains by making the standard finite memory approximation.

3.6 Proof of Principle

To demonstrate the effectiveness of TEMPO it is now used to numerically calculate the dynamics of an open TLS. The results presented here are intended simply to show that there is indeed significant computational improvement upon existing methods that are capable of producing the same results. Application to regimes that were not previously accessible with existing methods is found in Chap. 4. The implementation of TEMPO used in this section was written in Python and can be accessed through the link provided through Ref. [19] (see also the 'Digital Outputs Access Statement' at the start of this thesis). Specifically, this implementation is designed to calculate the dynamics of a given open system with abelian coupling to a Gaussian, bosonic environment, and with the option of using the finite memory approximation described in Sect. 3.4. The truncation of a given bond dimension of the MPS that is being stored is determined by predefining a maximum ratio, ϵ, of the largest singular value found when making an SVD to the smallest one that should be retained after truncation. For $\epsilon = 1$ only one singular value is retained and the resulting bond dimensions of the MPS are truncated to $\lambda = 1$, while $\epsilon = 0$ is equivalent to making no truncation of singular values. Note that by setting $\epsilon = 0$, such that all singular values are retained while performing contractions, TEMPO and QUAPI are functionally identical. It is not clear a priori what the physical significance of the singular values is in the context of TEMPO, so a general justification for the truncation of bond dimensions cannot be given. However, the results presented in this section provide numerical evidence that such truncation is effective, with the results of calculations converging rapidly as the limit $\epsilon \to 0$ is taken.

The model used for testing is that of a degenerate TLS experiencing unitary driving and coupling to a HB via its eigenstates. The Hamiltonian for the overall closed system is that of the spin-1/2 spin-boson model,

$$H = \Omega x_\sigma + \left(p_\sigma - \frac{1}{2}\right) \sum_k g_k (b_k + b_k^\dagger) + \sum_k \omega_k b_k^\dagger b_k. \tag{3.100}$$

The coupling operator $x = \left(p_\sigma - \frac{1}{2}\right)$ is the difference in projection operators onto the excited and ground states of the TLS. The temperature of the HB is taken to be $\beta = 0.1\Omega$ and the SD is assumed to be Ohmic, of the form Eq. (2.44) with $\nu = 1$, with exponential cutoff defined by $\omega_c = 3.25\Omega$, and dimensionless coupling strength $\alpha = 0.1$. This model and parameter set were used originally to demonstrate the numerical evaluation of a discretised Feynman-Vernon IF using QUAPI [3, 17].

The dynamics of $\langle p_\sigma(t) \rangle$ calculated using QUAPI and TEMPO are plotted in Fig. 3.2a. While there have been numerous developments and improvements of the original QUAPI technique [20–23], only the standard technique presented in Ref. [3] is considered here which uses only a discretisation and the finite memory approximation to perform calculations. The black stars show data calculated using QUAPI, whose exponential scaling of storage requirements mean that a finite memory approximation, $K = 10$, has to be used. It was demonstrated in Ref. [3] that this value of K is sufficient for numerical convergence for this set of parameters. The coloured lines show results calculated using TEMPO for varying ϵ, whose efficiency is such that a finite memory approximation is not required. As expected, the dynamics calculated using TEMPO converge to the same dynamics as calculated using QUAPI as ϵ is decreased. In terms of computing time, generation of the converged results in Fig. 3.2a using TEMPO (red line) and QUAPI (black stars) took ~13 s and ~12 min, respectively, on a dual core laptop computer with 16 GB RAM. The total number of scalar numbers N_{tot} required to store the MPS in TEMPO and the ADT in QUAPI for varying K are plotted in Fig. 3.2b. The truncation of bond dimensions in this case is fixed at $\epsilon = 10^{-3.5}$. The storage of the ADT in QUAPI scales exponentially, $N_{tot} = 4^K$, shown by the orange line. In contrast, the storage of the MPS in TEMPO only grows linearly $N_{tot} \propto K$, shown by the blue line and made clear in the inset. The vast improvement in storage requirements gained by using TEMPO can be seen by the fact that N_{tot} for TEMPO with $K = 100$ (the case of no memory cutoff in Fig. 3.2a) is less than N_{tot} for QUAPI with $K = 10$.

When applied to investigate the quite generic model of a TLS with abelian coupling given by Eq. (3.100), it is not yet clear what the limitations of TEMPO are. Variants

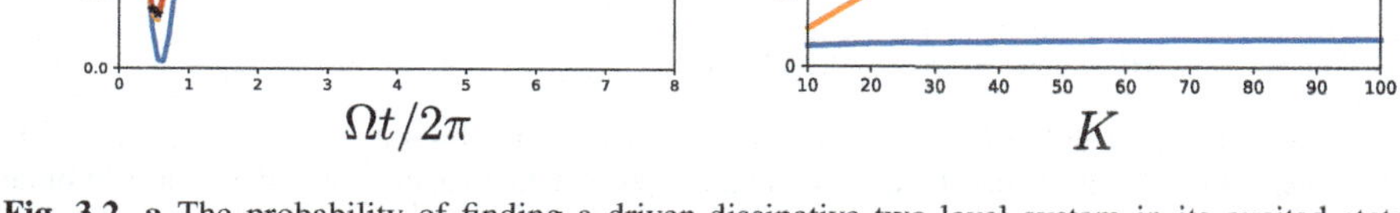

Fig. 3.2 **a** The probability of finding a driven-dissipative two-level system in its excited state calculated using TEMPO and QUAPI. The blue, orange, green, and red lines depict the data calculated by using TEMPO with $\epsilon = 10^{-1}, 10^{-1.5}, 10^{-2.5}, 10^{-3.5}$ respectively with no finite memory approximation. The black stars depict data taken using QUAPI with $K = 10$. Parameters used are $\beta = 0.1\Omega$, $\omega_c = 3.25\Omega$, $\alpha = 0.1$. **b** The logarithm of N_{tot} against K using the same system parameters as in panel (**a**). The blue line is N_{tot} required for storage of the MPS in TEMPO with $\epsilon = 10^{-3.5}$ and the orange line is N_{tot} required to store the ADT in QUAPI. The inset shows N_{tot} for TEMPO plotted on a linear scale

of the model will be investigated further in Chap. 4 where it is found that TEMPO is effective for all parameter regimes that are explored. In particular, no simple physical picture has yet emerged that could indicate for what parameters the truncation of singular values becomes unfeasible. This is partly because such a parameter regime has yet to be found. One limitation of TEMPO that is encountered when going beyond models like Eq. (3.100) is that of the size of the tensors that make up the uncontracted TN representing a given problem. For a general n-level system with non-abelian coupling to an environment it was shown in Sect. 3.3 that the largest tensors in the network are made up of n^{16} complex scalars, while for abelian couplings this number can be reduced to n^8. Assuming a single real number requires 4 bytes of memory then the memory required to store a single tensor of the network for a TLS ($n = 2$) with nonabelian coupling is $\sim$17 GB, making this situation impossible to address on a typical laptop computer and generally difficult even using more powerful machines. For abelian couplings the size of a single tensor of a network becomes similarly large, $\sim$17 GB, for $n = 8$.

Finally, the effectiveness of the improved finite memory approximation introduced in Sect. 3.5 is now demonstrated using QUAPI, for which it was originally developed. The improved finite memory approximation can equally well be implemented within the framework of TEMPO, but the effectiveness of the approximation can be better assessed without having to consider the additional errors induced by discarding singular values as would occur when using TEMPO. The model considered is the same as in Eq. (3.100) but using a superohmic SD, of the form Eq. (2.44) with $\nu = 3$ and an exponential cutoff with cutoff frequency $\omega_c = 5\Omega$. It was shown in Sect. 3.5 that this type of SD is expected to result in the most severe error when using the standard finite memory approximation. The temperature of the HB has been taken to $\beta^{-1} = 0$ and the dimensionless coupling is $\alpha = 0.35$. The results of calculating the excited state probability using the finite memory cutoff and the improved finite memory cutoff with QUAPI are shown in Fig. 3.3a and b using $K = 11$ and $K = 6$ respectively. In both cases the use of the improved finite memory

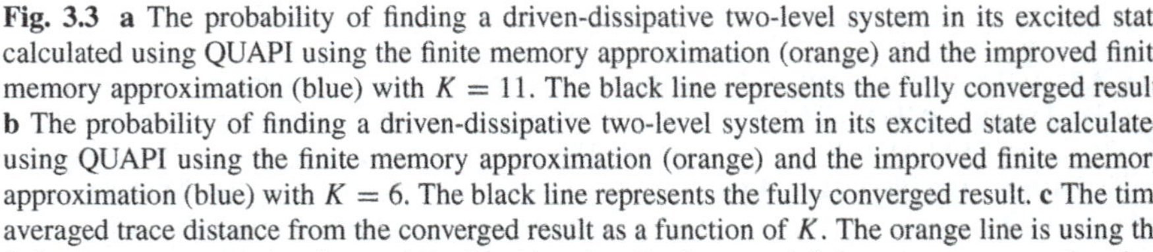

Fig. 3.3 **a** The probability of finding a driven-dissipative two-level system in its excited state calculated using QUAPI using the finite memory approximation (orange) and the improved finite memory approximation (blue) with $K = 11$. The black line represents the fully converged result. **b** The probability of finding a driven-dissipative two-level system in its excited state calculated using QUAPI using the finite memory approximation (orange) and the improved finite memory approximation (blue) with $K = 6$. The black line represents the fully converged result. **c** The time averaged trace distance from the converged result as a function of K. The orange line is using the finite memory approximation and the blue is using the improved finite memory approximation. This figure is reproduced with permission from Ref. [18]

cutoff produces results much closer to the fully converged result, and for $K = 6$ the finite memory cutoff not only is less converged but results in probabilities less than zero and greater than one which is unphysical. The approach to convergence using the two types of finite memory approximation is shown in Fig. 3.3c. The approach to convergence was measured by calculating the trace distance, defined in Eq. (2.92), between the unconverged RDM, $\rho_1(t)$ and the converged RDM, $\rho_2(t)$, at all points in time and taking the average. The improved finite memory approximation is closer to the converged result over full range of K shown, although the rate of approach to convergence with K is approximately the same as that of the standard finite memory cutoff.

3.7 Summary

The first result given in this chapter, in Sect. 3.1, was that the application of the time ordering operator in any situation can represented diagrammatically as a TN. The ubiquity of time ordering in quantum mechanics makes this a significant result on its own. The further methodology presented in Sect. 3.2 showed that this TN representation of time ordering has particular use when considering time ordered IFs of open quantum systems. Though only introductory examples of the TNs used to represent such problems were given, the generality of the connection made between open quantum systems and TNs opens the door to further research. It was then found TNs representing open quantum systems can be contracted efficiently using well known matrix product techniques in Sect. 3.3. The finite memory approximations discussed in Sects. 3.4 and 3.5 allowed reduction of the these TNs, making their contraction yet more feasible. Finally, the effectiveness of the methodology presented in this chapter was demonstrated using a toy model for an open quantum system in Sect. 3.6. Comparison to methods previously used to solve this model revealed an substantial improvement of efficiency.

References

1. Diósi L, Ferialdi L (2014) General non-Markovian structure of gaussian master and stochastic schrödinger equations. Phys Rev Lett 113:200403
2. Ferialdi L (2016) Exact closed master equation for gaussian non-Markovian dynamics. Phys Rev Lett 116:120402
3. Makri N, Makarov DE (1995) Tensor propagator for iterative quantum time evolution of reduced density matrices. I. Theory. J Chem Phys 102(11):4600
4. Macedo HD, Oliveira JN (2013) Typing linear algebra: a biproduct-oriented approach. Sci Comput Progr 78(11):2160
5. Arrighi P, Patricot C (2004) On quantum operations as quantum states. Ann Phys 311(1):26
6. Gill TL (2017) The Feynman-Dyson view. J Phys Conf Ser 845:012023
7. Gill TL, Zachary WW (2011) Feynman operator calculus: the constructive theory. Expos Math 29(2):165

8. Feynman RP (1951) An operator calculus having applications in quantum electrodynamics. Phys Rev 84(1):108
9. Jefferies B, Johnson GW, Kim BS (2006) Feynman's operational calculi: methods for iterative disentangling. Acta Applicandae Mathematicae 92(3):293–309
10. Pollock FA, Rodríguez-Rosario C, Frauenheim T, Paternostro M, Modi K (2018) Non-Markovian quantum processes: complete framework and efficient characterization. Phys Rev A 97:012127
11. Jørgensen MR, Pollock FA (2019) Exploiting the causal tensor network structure of quantum processes to efficiently simulate non-Markovian path integrals. Phys Rev Lett 123:240602
12. Chiribella G, D'Ariano GM, Perinotti P (2008) Quantum circuit architecture. Phys Rev Lett 101:060401
13. Cotler J, Jian C-M, Qi X-L, Wilczek F (2018) Superdensity operators for spacetime quantum mechanics. J High Energy Phys 9:2018
14. Grabert H, Schramm P, Ingold G-L (1988) Quantum Brownian motion: the functional integral approach. Phys Rep 168(3):115–207
15. Strathearn A, Kirton P, Kilda D, Keeling J, Lovett BW (2018) Efficient non-Markovian quantum dynamics using time-evolving matrix product operators. Nat Commun 9:3322
16. Makri N (1995) Numerical path integral techniques for long time dynamics of quantum dissipative systems. J Math Phys 36(5):2430
17. Makri N, Makarov DE (1995) Tensor propagator for iterative quantum time evolution of reduced density matrices. II. Numerical methodology. J Chem Phys 102(11):4611
18. Strathearn A, Lovett BW, Kirton P (2017) Efficient real-time path integrals for non-Markovian spin-boson models. New J Phys 19(9):093009
19. Strathearn A, Kilda D (2018) Python implementation of TEMPO. https://doi.org/10.5281/zenodo.1322407
20. Sim E (2001) Quantum dynamics for a system coupled to slow baths: on-the-fly filtered propagator method. J Chem Phys 115(10):4450
21. Sim E, Makri N (1996) Tensor propagator with weight-selected paths for quantum dissipative dynamics with long-memory kernels. Chem Phys Lett 249(3–4):224
22. Sim E, Makri N (1997) Filtered propagator functional for iterative dynamics of quantum dissipative systems. Comput Phys Commun 99(2–3):335
23. Makri N (2017) Iterative blip-summed path integral for quantum dynamics in strongly dissipative environments. J Chem Phys 146(13):134101

Chapter 4
Results

In this chapter, the methods described in the previous chapter for numerically evaluating time ordered expressions are applied to study three simple models of open quantum systems that exhibit non-Markovian phenomena. First, the localisation phase transition in the Spin-Boson model is studied in Sect. 4.1, where it is shown that the location of the transition can be identified numerically to produce results consistent with existing literature. Localisation is non-Markovian by nature since it involves infinite ranged memory effects. In Sect. 4.2 the phenomenon of environment mediated interactions is investigated in a model of two spins interacting with a common environment. In this model the propagation of environmental excitations between the spins is nontrivial non-Markovian behaviour that is not possible to capture using existing methods, and the results presented in Sect. 4.2 are the first to confirm that the system behaves in an intuitive manner. The fluorescence spectrum of a driven-dissipative quantum dot is then studied in Sect. 4.3. Such spectra are observable in experiment and have been seen to exhibit a phonon sideband, a feature arising from the non-Markovian behaviour of the vibrational degrees of freedom in the dot. This sideband is seen in the spectra calculated in Sect. 4.3 and the behaviour of the spectra over a wide range of parameter space is investigated.

The three models studied in this chapter are highly idealised and yet each of them present some form of difficulty. The body of knowledge surrounding these models has been built up using a plethora of analytic and numerical methods, typically tailored to provide answers to specific questions about each model. Using the methods outlined in Chap. 3, numerical solution of all three models is straightforward.

A. Strathearn, *Modelling Non-Markovian Quantum Systems Using Tensor Networks*,
Springer Theses, https://doi.org/10.1007/978-3-030-54975-6_4

4.1 Localisation

The Spin-Boson model (SBM) [1] describes a single spin-S degree of freedom coupled to a HB. The Hamiltonian for the model is

$$H = \Omega S_x + \sum_k S_z(g_k b_k + g_k^* b_k^\dagger) + \omega_k b_k^\dagger b_k, \tag{4.1}$$

where S_ν is spin operator in the ν direction, b_k is the annihilation operator for the k'th mode of the HB, with frequency ω_k and coupling g_k. The spin could, for example, represent the intrinsic spin of a particle, a magnetic dipole, or simply angular momentum. Aside from providing a simple toy model with which theoretical aspects of open systems can be investigated, the SBM can be realised experimentally using superconducting qubits [2–5].

When the SD of the HB in the SBM is Ohmic,

$$J(\omega) = 2\alpha\omega e^{-\frac{\omega}{\omega_c}}, \tag{4.2}$$

the existence of two distinct phases of the model is known [1]. The transition is in the Kosterlitz-Thouless universality class [6] and is parameterised by the coupling strength, α, with the transition from a delocalised phase to a localised phase occurring as α is increased through a critical value α_c [1, 7]. A similar localisation transition, though not of the Kosterlitz-Thouless type, is also known to occur when the SD is subohmic [8–10]. The phase transition in the SBM can be studied both in an equilibrium setting [11], in which case there is symmetry breaking in the collective ground state of Eq. (4.1), or in dissipative setting [12], in which case symmetry breaking manifests in the steady state of the spin alone. The latter of these setting shall be considered here. The Hamiltonian of the SBM is invariant under $S_z \to -S_z$ and $b_k \to -b_k$ which, upon tracing out the HB, translates to an invariance of time-nonlocal Liouvillian for the spin under $S_z \to -S_z$. For $\alpha < \alpha_c$ the spin reaches a steady state which respects this symmetry, $\lim_{t\to\infty}\langle S_z(t)\rangle = 0$, while for $\alpha > \alpha_c$ the system localises around its initial state, $\lim_{t\to\infty}\langle S_z(t)\rangle \propto \langle S_z(0)\rangle$, breaking the symmetry.

For the spin-1/2 SBM with a zero temperature HB the critical coupling is known to be $\alpha_c = 1 + \mathcal{O}(\Omega/\omega_c)$ [1, 6, 14]. The dynamics of $\langle S_z(t)\rangle$ for this case are calculated in the coupling range $0 < \alpha < 1.5$, with $\Omega/\omega_c = 0.2$, and displayed in Fig. 4.1a. To achieve numerical convergence the discretisation timestep used is $\Omega\Delta = 0.33$. The finite memory approximation described in Sect. 3.4 is employed with $K = 200$. The truncation of bond dimensions of the MPS is set by the ratio of the smallest singular value to be retained after truncation to the largest singular, $\epsilon = 10^{-12}$. The hard cutoff time, $\tau_c = \Delta K$, is marked with a vertical black line in Fig. 4.1a. In the weak coupling regime, $0 < \alpha < 0.5$, the dynamics display increasingly damped oscillations around $\langle S_z(t)\rangle = 0$ at a frequency $\sim\Omega$. At $\alpha \approx 0.5$ these oscillations become overdamped, such that the spin undergoes pure relaxation to its symmetric steady state. In the

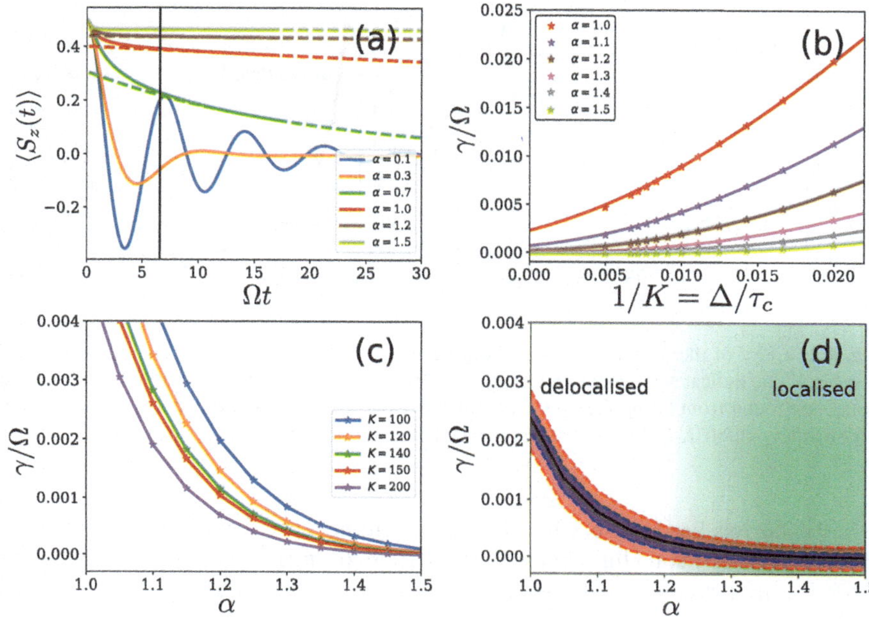

Fig. 4.1 **a** Plot of the dynamics of the z-component of a spin-1/2 with varying coupling strength. Dashed lines indicate the exponential decay found from fitting to the data. The cutoff frequency used was $\Omega/\omega_c = 0.2$. **b** The decay rate found from fitting the asymptotic dynamics of $\langle S_z \rangle$ to an exponential decay varying with memory cutoff K. Stars indicate the values of γ found from fitting to an exponential and lines indicate the fitting of these extracted decay rates to a polynomial in $1/K$. **c** The same data as in panel **b** plotted to show how the extrapolated decay rates vary with coupling for a given value of K. **d** Variation of decay rate with coupling extrapolated to $K \to \infty$ with blue and red regions indicating 68% and 95% confidence intervals respectively. This figure is reproduced with permission from Ref. [13]

limit $\omega_c \to \infty$, this underdamped to overdamped crossover happens at precisely $\alpha = 0.5$, which can be inferred exactly [15] through a number of methods. As the coupling is increased beyond this crossover, the relaxation time of the spin decreases until it appears to reach a symmetry broken steady state for $\alpha > 1$. Localisation is a temporally infinite ranged effect since the steady state is dependent on the initial state, so the use of the finite memory approximation means that the spin cannot truly localise. This means that the localised steady state implied for $\alpha = 1, 1.5$ in Fig. 4.1a is in fact not stable and does decay very slowly to $\langle S_z(t) \rangle = 0$.

In order to numerically identify the critical coupling, α_c, the behaviour of the apparent localisation for $\alpha > 1$ is analysed as a function of the cutoff K. Rather that localising, the asymptotic dynamics of the spin are found to be of exponential form, $\langle S_z(t) \rangle \sim e^{-\gamma t}$. A fitting to this exponential form are shown as dashed lines in Fig. 4.1a. By performing this fitting for various values of K and α, the functional dependence of the asymptotic exponential decay rate, $\gamma = \gamma(K, \alpha)$, is extracted and displayed in Fig. 4.1b, c. It is expected that extrapolating to $K \to \infty$ should result in

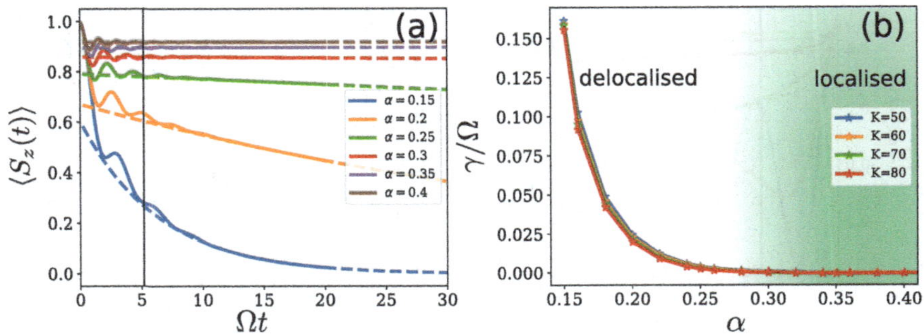

Fig. 4.2 **a** Plot of the dynamics of the z-component of a spin-1 with varying coupling strength. The black line indicates the memory cutoff time. The cutoff frequency used was $\omega_c = 5\Omega$. **b** The decay rate found from fitting the asymptotic dynamics of $\langle S_z(t) \rangle$ to an exponential decay varying with memory cutoff K. This figure is reproduced with permission from Ref. [13]

$\gamma = 0$ in the localised phase, which is found by fitting γ to a second order polynomial in K, indicated by solid lines in Fig. 4.1b, c. These fittings suggest a critical coupling $\alpha_c \approx 1.2$, in agreement with the analytic result $\alpha_c = 1 + \mathcal{O}(\Omega/\omega_c)$ [1, 6, 14]. To more precisely identify α_c, the numerical error in the extrapolated values of γ is estimated and displayed in Fig. 4.1d. The black line indicates the extrapolated decay as a function of α, while the red and blue lines indicate intervals of 68% and 95% confidence respectively. This analysis implies a more accurate estimate of the critical coupling to be $\alpha_c \approx 1.25$. A yet more accurate estimate for α_c could be obtained by increasing the maximum value of K used in numerical calculations, though the maximum value used here $K = 200$ is sufficient to demonstrate effectiveness of the method and to confirm agreement with analytics.

The efficiency of the techniques used allows for consideration of the SBM with a larger spin. The dynamics of $\langle S_z(t) \rangle$ for the spin-1 SBM model are shown in Fig. 4.2a. The black line indicates the memory cutoff time $\tau_c = K\Delta$. In this case coherent oscillations are present for all α, such that an underdamped to overdamped crossover point cannot be identified as in the spin-1/2 case. The asymptotic dynamics can however be fitted to exponential decay again the results of which are shown in Fig. 4.2b. Without performing extrapolation to $K \to \infty$ or error analysis, the transition to a localised phase becomes apparent at $\alpha_c \approx 0.25$, in agreement with renormalisation group calculations [16, 17] but in disagreement with calculations performed using a variational ground state ansatz [18].

Before concluding this section, a brief analysis of the computational resources that were required to perform the calculations displayed in Fig. 4.1a is now given. It was found that the optimal protocol to truncate bond dimensions was to sweep along the MPS in one direction performing truncations and then to sweep back truncating all bonds again, which resulted in the largest bond dimensions being the center of the MPS. The total storage of the MPS at the end of a calculation, N_{tot}, along with the maximum bond dimension, λ_{max}, are plotted in Fig. 4.3a as a function of coupling strength. It is seen the both have a maximum around $\alpha = 0.5$,

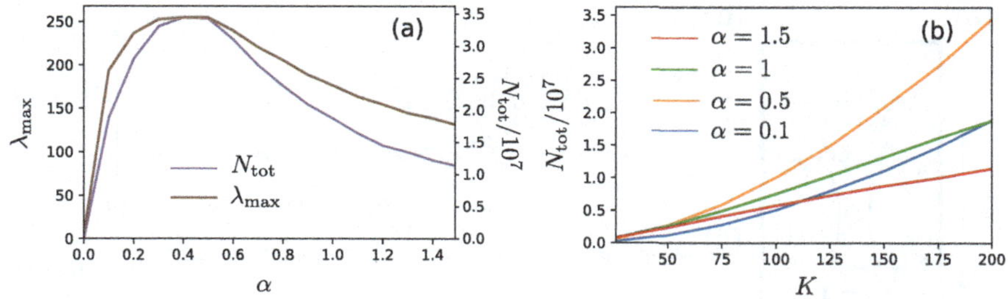

Fig. 4.3 **a** The total size, N_{tot}, and maximum bond dimension, λ_{max}, for the MPS at the end of performing the calculations for Fig. 4.1a as a function of coupling strength. **b** The total size, N_{tot}, of the MPS at the end of performing the calculations for Fig. 4.1a for various coupling strengths as a function of the memory cutoff K. This figure is reproduced with permission from Ref. [13]

where the dynamics of $\langle S_z(t) \rangle$ crossover from underdamped to overdamped, rather than around the localisation transition point. This is somewhat surprising since one might expect that the transition to a phase in which time-nonlocality becomes infinite ranged would be more computationally demanding, but in fact there are no significant features around the transition point in Fig. 4.3a. The total size of the MPS for various coupling strengths as a function of K is shown in Fig. 4.3b. Again rather surprisingly, the storage is seen to scale better with K for larger coupling. For lower couplings N_{tot} is seen to increase polynomially with K which becomes approximately linear as coupling is increased.

4.2 Environment Mediated Interactions

Another phenomenon that is intrinsically non-Markovian is that of an environment mediated interaction. If an open quantum consists of spatially separated subsystems immersed in a common environment, perturbation of the environment by a single subsystem can lead to backaction of the environment on a different subsystem at a later time. In terms of the degrees of freedom of the subsystems alone, this can be thought of as an effective retarded interaction between subsystems. Such effects have been seen in experiments with artificial atoms [19]. A toy model in which this effect can be seen is a model of a pair of interacting spins which couple to a single HB, shown schematically in Fig. 4.4a. The Hamiltonian for this model is,

$$H = \Omega \mathbf{S}_a \cdot \mathbf{S}_b + \sum_{\nu=a,b} \sum_{\mathbf{k}} S_{z,\nu}(g_{\mathbf{k},\nu} b_{\mathbf{k}} + g^*_{\mathbf{k},\nu} b^\dagger_{\mathbf{k}}) + \omega_{\mathbf{k}} b^\dagger_{\mathbf{k}} b_{\mathbf{k}}, \tag{4.3}$$

where $\mathbf{S}_\nu = (S_{x,\nu}, S_{y,\nu}, S_{z,\nu})$ is the spin vector for the νth spin. The spatial separation of the spins is accounted for through the spatially dependent coupling to the HB, $g_{\mathbf{k},\nu} = |g_{\mathbf{k}}| e^{-i\mathbf{k}\cdot\mathbf{r}_\nu}$, where $\mathbf{r}_\nu$ is the position vector of the νth spin. This model could describe atoms placed at different positions along a 1D waveguide [20]. Assuming a

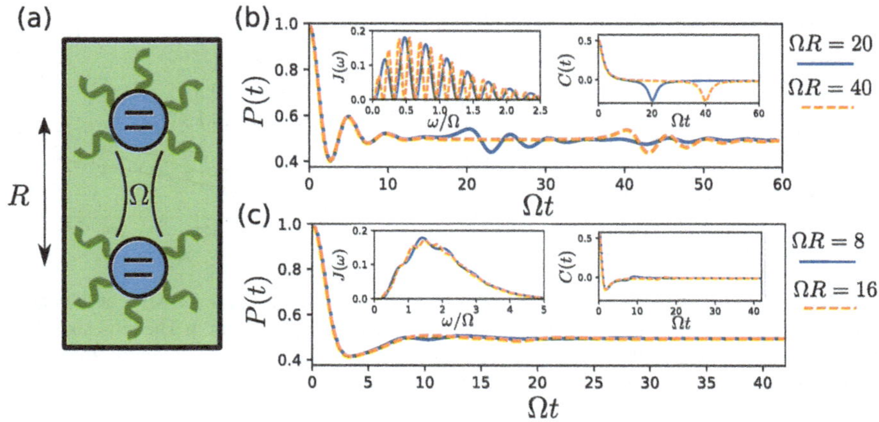

Fig. 4.4 **a** A schematic of two two-level systems interacting with each other and with a common environment. **b** The probability of finding the spins in an anti-aligned state as a function of time for a one dimensional environment, $D = 1$. Blue (orange) lines indicate a spatial separation of $\Omega R = 10\,(\Omega R = 20)$. The left inset shows the effective spectral density and the right inset shows the corresponding response function. Other parameters used where $\Omega\beta = 2, \alpha = 2, \omega_c/\Omega = 0.5$. **c** The probability of finding the spins in an anti-aligned state as a function of time for a three dimensional environment, $D = 3$. Blue (orange) lines indicate a spatial separation of $\Omega R = 8$ ($\Omega R = 16$). The left inset shows the effective spectral density and the right inset shows the corresponding response function. Other parameters used where $\Omega\beta = 2$, $\alpha = 1$, $\omega_c/\Omega = 0.5$. This figure is reproduced with permission from Ref. [13]

deformation-type coupling $|g_{\mathbf{k}}| \propto \sqrt{\omega_{\mathbf{k}}}$ [21] in D spatial dimensions, the SD of the HB can be written

$$J(\omega) = \sum_{\mathbf{k}} |g_{\mathbf{k}}|^2 = \frac{\alpha}{2} \frac{\omega^D}{\omega_c^{D-1}} e^{-\frac{\omega}{\omega_c}}. \tag{4.4}$$

The model described by Eq. (4.3) has an integral of motion $[H, S_{z,a} + S_{z,b}] = 0$. It is assumed that the spins are both spin-1/2, such that the dimension of the open system is $d = 4$. There are then three decoupled subspaces of the system: a two dimensional subspace with $S_{z,a} + S_{z,b} = 0$, in which the spins are anti-aligned, and two one dimensional subspaces with $S_{z,a} + S_{z,b} = \pm 1$. The dynamics of the one dimensional subspaces are trivial, and so only the two dimensional subspace need be considered. The part of the Hamiltonian in Eq. (4.3) responsible for the interaction can be separated into a piece $\propto S_{z,a} + S_{z,b}$ and a piece $\propto S_{z,a} - S_{z,b}$, both of which have the two the anti-aligned spin states as eigenvectors. The eigenvalues corresponding to the term $\propto S_{z,a} + S_{z,b}$ are both 0 and so it can be discarded. The term $\propto S_{z,a} - S_{z,b}$ has eigenvalues ± 1, so when restricting to the two dimensional subspace spanned by the anti-aligned spin states it has the same action as the operator $2S_z$ acting on a single spin-1/2, whose spin up and down states are the two anti-aligned spin states. Similarly, the operator $\mathbf{S}_a \cdot \mathbf{S}_b$ has the same action on these states as S_x does on a spin-1/2.

Thus, when restricting attention to the two anti-aligned spin states, Eq. (4.3) can be mapped onto the Hamiltonian for a single spin-1/2 interacting with a HB,

$$H = \Omega S_x + S_z \sum_{\mathbf{k}} 2i|g_{\mathbf{k}}| \sin\left(\frac{\mathbf{k}\cdot\mathbf{R}}{2}\right)(b_{\mathbf{k}} - b_{\mathbf{k}}^{\dagger}) + \omega_{\mathbf{k}} b_{\mathbf{k}'}^{\dagger} b_{\mathbf{k}}, \tag{4.5}$$

where $\mathbf{R} = \mathbf{r}_a - \mathbf{r}_b$. This is the Hamiltonian for a spin-1/2 SBM with a HB with SD given by

$$\tilde{J}(\omega) = 4\sum_{\mathbf{k}} |g_{\mathbf{k}}|^2(1 - \cos(\mathbf{k}\cdot\mathbf{R})) \tag{4.6}$$

Assuming a spatially isotropic environment, $\omega_{\mathbf{k}} = c|\mathbf{k}|$, where $c = 1$ is the speed of environmental excitations, the summation in Eq. (4.6) becomes an angular averaging in D dimensions and evaluates to

$$\tilde{J}(\omega) = 2J(\omega)(1 - F_D(\omega R)). \tag{4.7}$$

where $R = |\mathbf{R}|$ and $J(\omega)$ is as defined in Eq. (4.4). The function $F_D(R, \omega)$ encodes how collective excitations in the HB propagate between spins and is given by

$$F_D(x) = \begin{cases} \cos(x) & D = 1 \\ b_0(x) & D = 2 \\ \mathrm{sinc}(x) & D = 3 \end{cases}, \tag{4.8}$$

where $b_0(x)$ is a Bessel function. For $D > 1$ there is loss of intensity of environmental excitations as they propagate through multidimensional space such that $\lim_{R\to\infty} F_D(\omega R) = 0$. In this limit the environment has an effectively independent influence on each of the spins, and if the influence of the bare SD, $J(\omega)$, is effectively Markovian for each spin then analysis of the model is possible using methods outlined in Sect. 2.1.2. Conversely, in the limit of vanishing separation, $R \to 0$, there is a negligible relative phase between the coupling of each spin to the HB in Eq. (4.3) such that the two dimensional subspace being considered effectively decouples from the HB. For small but finite R, the regime where the function $F_D(\omega R)$ can be expanded in R to get $\tilde{J}(\omega) \propto \omega^2 R^2 J(\omega)$, the use of the polaron frame [22], as outlined in Sect. 2.1.3.2, or QUAPI [23] allows for accurate modelling of the system. Therefore in both the limits of both large and small R the model is amenable to existing methods, so the intermediate regime, $R\omega_c \sim 1$, is considered here.

It is assumed the pair of spins are initialised at $t = 0$ in one of their two anti-aligned states with probability one. The probability of finding them in this state as a function of time, $P(t)$, is calculated numerically and plotted in Fig. 4.4b, c. In Fig. 4.4b the dimension of the environment has been taken as $D = 1$ and the dynamics of $P(t)$ are plotted for two different separations, $\Omega R = 10$ and $\Omega R = 20$. There is an initial oscillatory decay of probability around the steady state value $\lim_{t\to\infty} P(t) = 1/2$ until $t \approx R$ where there is a revival of probability which then oscillates and decays to the steady state value. For both values of R the profile of the revival at $t \approx R$ is the same, which is expected for $D = 1$ since there is no loss of intensity of environmental excitations as they propagate between the two spins. In Fig. 4.4b the dimension of the environment has been taken as $D = 3$ and the dynamics of $P(t)$ are plotted for two different separations, $\Omega R = 8$ and $\Omega R = 16$. In this case there are once again revivals at $t \approx R$ but the revival for $\Omega R = 16$ is much less pronounced than the revival for $\Omega R = 8$ due to loss of intensity as environmental excitations propagate through three dimensional space. The insets to Fig. 4.4b, c show the effective SDs and respective response functions of the environment. In each case there are modulations in the SD with period R^{-1} which manifest as a feature in the response function at $t = R$. For $D = 3$ the effect of the loss of intensity of environmental excitations can be seen as a lower amplitude of modulation in the effective spectral density for larger R and a resulting smaller feature in the response function.

4.3 Driven-Dissipative Quantum Dot

A quantum dot is a small piece of semiconducting material with a distinctive quantisation of its electronic states. Quantum dots are of interest in the development of emerging quantum technologies, having been demonstrated as strong candidates for scalable on-demand single photon sources [24–26] as well as being capable of serving as components of a quantum computer [26–29]. Basic properties of quantum dots are readily controlled and accessible in experiment. For instance, quantum dots of a given size can be synthesised to produce a desired splitting of the quantised energy levels, which in turn determines the frequency of light they emit [30]. The vibrational, or phononic, degrees of freedom intrinsic to quantum dots act as an environment to their electronic, or excitonic, states with measurable effect [31]. This makes them a good platform to test and develop mathematical models of few-level open quantum systems. In particular, the coupling of excitonic states to phononic degrees of freedom has been observed to result in a phonon sideband in the fluorescence spectra of quantum dots [32], a phenomenon whose description is known to require methods that go beyond Markovian weak coupling theory [33, 34]. Understanding exciton-phonon coupling is also of practical interest, since such coupling can result in degradation of the quality of emitted single photons [35, 36], for example.

The model for a quantum dot considered here consists of a single excitonic state with energy ω_0 whose transition is driven by an electromagnetic field, into which the exciton can also radiatively decay, and which couples to the phonon degrees of freedom intrinsic to the quantum dot. A schematic of this model is shown in Fig. 4.5a. The ground state and single excitonic state are represented as the spin down and spin up states of a single spin-1/2 and both the electromagnetic field and the phonons in the quantum dot are modelled as HBs. It is assumed that dissipative influence of the driving electromagnetic field is weak enough that the RSM approximation can be used to describe it and that the excitonic transition is being driven resonantly. The effective Hamiltonian that describes the excitonic and phononic degrees of freedom

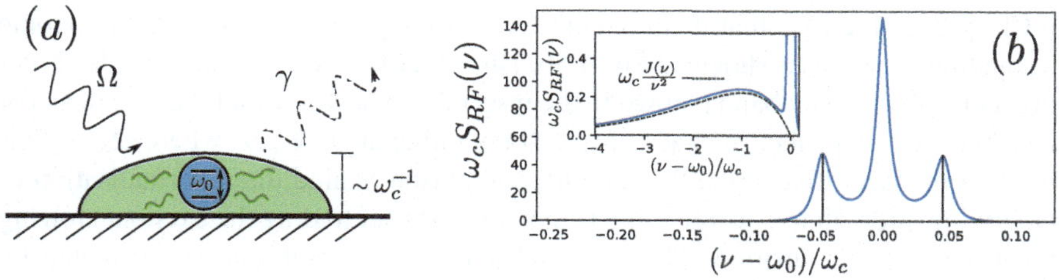

Fig. 4.5 **a** A schematic of a driven-dissipative quantum dot. **b** An illustrative example of a numerically calculated fluorescence spectrum of a driven-dissipative quantum dot. Black solid lines indicate the position of the maxima of the sidepeaks. The inset shows the same spectrum on a different scale to show the phonon sideband. The black dashed line indicates the analytic prediction for the sideband in the limit of zero drive. Parameters used are $\Omega = 0.05\omega_c$, $\gamma = 0.01\omega_c$, and $\alpha = 0.1$

is then

$$H = \Omega S_x + \left(S_z + \frac{1}{2}\right)\sum_{\mathbf{k}} g_{\mathbf{k}}(b_{\mathbf{k}} + b_{\mathbf{k}}^\dagger) + \sum_{\mathbf{k}} \omega_{\mathbf{k}} b_{\mathbf{k}}^\dagger b_{\mathbf{k}}, \tag{4.9}$$

where Ω is the driving strength. The weak dissipative influence of the electromagnetic field is accounted for through a dissipator, $\gamma\mathcal{D}(S^-)$, where S^- is the lowering operator for the spin. To better isolate the effects of the interplay between phonon induced dissipation and electromagnetic driving it is assumed that the phonons are initially in their collective ground state. It is also assumed that the exciton couples to isotropically dispersed phonons, $\omega_{\mathbf{k}} = c|\mathbf{k}|$, where $c = 1$ is the phonon propagation speed, in a three dimensional lattice through a deformation type coupling, $g_{\mathbf{k}} \sim \sqrt{\omega_{\mathbf{k}}}$. The SD then takes the form

$$J(\omega) = \alpha \frac{\omega^3}{\omega_c^2} e^{-\frac{\omega}{\omega_c}}, \tag{4.10}$$

where the cutoff frequency, ω_c, is set by the spatial extent of the quantum dot and α is the dimensionless coupling strength.

A useful and experimentally accessible quantity for assessing the behaviour of a quantum dot is its incoherent resonance fluorescence spectrum

$$S_{RF}(\omega) = \int_{-\infty}^{\infty} \lim_{t\to\infty} \left(\langle S^+(t+\tau)S^-(t)\rangle - \langle S^+(t+\tau)\rangle\langle S^-(t)\rangle\right) e^{-i\omega\tau} d\tau, \tag{4.11}$$

where S^+ is the raising operator of the spin. The fluorescence spectrum of the model described by Eq. (4.9) in the limit $\alpha = 0$ was discussed in Sect. 2.1.2.4, where the operators of the spin-1/2 being used here are related to those of the TLS in Sect. 2.1.2.4 by $S_x = x_\sigma$, $S_z + 1/2 = p_\sigma$, and $S^- = \sigma$. The defining feature of $S_{RF}(\omega)$ in this case is triplet of peaks at frequencies $\nu = \omega_0, \omega_0 \pm \Omega$. In the limit that there is no interaction between the electromagnetic field and exciton, $\Omega = \gamma = 0$, the model was solved exactly in Sect. 2.1.3.5 and $S_{RF}(\omega)$ was found to consist solely of a phonon sideband. A typical fluorescence spectrum, calculated using the methods in Chap. 3, in the case that there is both weak driving, $\Omega/\omega_c = 0.05$ and phononic dissipation, $\alpha = 0.1$, is shown in Fig. 4.5b. This spectrum exhibits both a triplet structure and a phonon sideband (shown in the inset). Note however that the side peaks of the triplet are not peaked at exactly $\omega_0 \pm \Omega$ but rather at $\omega_0 \pm \Omega_R$, where $\Omega_R < \Omega$ is the renormalised drive strength. The physical reason behind the renormalisation of the driving strength to a lower value is that phonons act to drag the exciton, slowing down the rate of transition. The renormalised drive strength can be estimated by transforming into the polaron frame. Defining

$$h = \left(S_z + \frac{1}{2}\right) \sum_k \frac{g_k}{\omega_k}(b_k^\dagger - b_k), \tag{4.12}$$

the transformation of the Hamiltonian Eq. (4.9) into the polaron frame is [37]

$$H_P = e^h H e^{-h} = \frac{\lambda_c}{2} S_z + \Omega(S^+ B^+ + S^- B^-) + \sum_k \omega_k b_k^\dagger b_k, \tag{4.13}$$

where λ_c is the reorganisation energy of the phonons and

$$B^\pm = \exp\left(\pm \sum_k \frac{g_k}{\omega_k}(b_k^\dagger - b_k)\right). \tag{4.14}$$

The first term in Eq. (4.13) represents a shift in the natural energy of the exciton due to the reconfiguration of the lattice around it and the second term is a nonlinear coupling of the exciton transition to the phonons. This can be thought of as an operationally renormalised driving of the exciton, whose expectation value taken with respect to the phonons gives an estimate of the scalar renormalised drive strength,

$$\Omega(S^+\langle B^+\rangle + S^-\langle B^-\rangle) = \Omega_R S_x, \tag{4.15}$$

with

$$\frac{\Omega_R}{\Omega} = \langle B^\pm \rangle = \exp\left(-\frac{1}{2}\int_0^\infty d\omega \frac{J(\omega)}{\omega^2} \coth\left(\frac{\beta\omega}{2}\right)\right). \tag{4.16}$$

For the parameters used in Fig. 4.5b this evaluates to $\Omega_R/\Omega = 0.95$, in qualitative agreement with the numerical results. The sideband shown in the inset of Fig. 4.5b is well described by the profile $\sim J(\nu)/\nu^2$ derived in Sect. 2.1.3.5, whose maximum is exactly at $\nu = \omega_c$ under the assumption of no driving and weak coupling to phonons.

As an aside, note that linearising the exciton-phonon coupling in Eq. (4.13) by expanding $B^\pm$ to first order in the lab frame coupling gives

$$\Omega(S^+ B^+ + S^- B^-) \approx \Omega S_x + S_y \sum_k i\Omega \frac{g_k}{\omega_k}(b_k^\dagger - b_k). \tag{4.17}$$

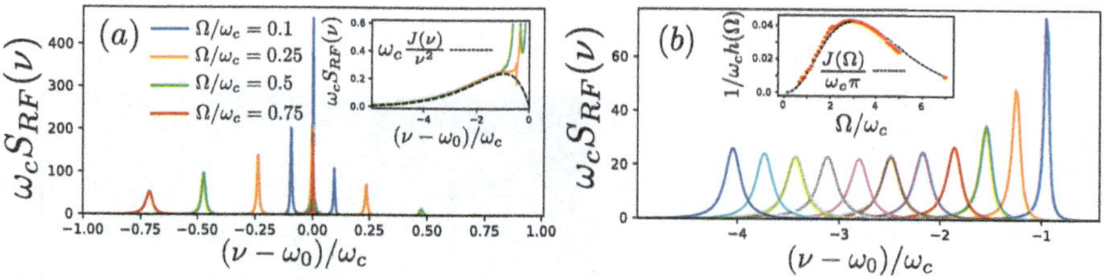

Fig. 4.6 **a** Fluorescence spectra for driving strengths of $\Omega/\omega_c = 0.1$ (blue), $\Omega/\omega_c = 0.25$ (orange), $\Omega/\omega_c = 0.5$ (green), and $\Omega/\omega_c = 0.75$ (red). The inset shows the spectra for $\Omega/\omega_c = 0.25, 0.5$ on a different scale and the black dashed line indicates the prediction for the sideband in the zero driving limit. Parameters used are $\gamma = 0.005\omega_c$ and $\alpha = 0.1$ **b** Negative frequency sidepeak of fluorescence spectra for driving strengths $4 < \Omega/\omega_c < 1$, increasing in increments of 0.3 from right to left. The inset shows the inverse height of the peak as a function of drive strength. The black dashed line is the spectral density divided by π. Parameters used are $\gamma = 0.001\omega_c$ and $\alpha = 0.1$

This describes interaction of the exciton with a HB with an effective SD, $J_P(\omega) = \Omega^2 J(\omega)/\omega^2$, which is broader, and thus more Markovian, than the SD in the lab frame as discussed in Sect. 2.1.3.2.

Fluorescence spectra for various driving strengths, $\Omega/\omega_c < 1$, are calculated and plotted in Fig. 4.6a. The asymmetry present in the height of the sidepeaks of the triplet is due to the fact that phonons assist mainly in the transition of the exciton from its upper dressed state to its lower dressed state which enhances emission into the lower frequency sidepeak. A weak coupling treatment of the phonons would result in incoherent excitation $\propto N(\beta)$ and spontaneous decay $\propto N(\beta) + 1$ of the upper dressed state, where $N(\beta)$ is the total expected phonon number. Since the phonons are initially in their ground state $\beta^{-1} = 0$ and $N(\beta) = 0$ so decay is preferred. The asymmetry becomes more pronounced as the drive strength is increased towards $\Omega/\omega_c = 1$, the regime in which the exciton transition and phonon relaxation occur over a similar timescales and for which the maxima of lower frequency sidepeak and the phonon sideband are expected to coincide. This can then be thought of as the regime in which the exciton transition is on resonance with the collective phonon transitions that cause decay into the lower dressed state of the exciton, explaining why the asymmetry is more pronounced as $\Omega/\omega_c = 1$ is approached. The inset of Fig. 4.6a shows that the phonon sideband and lower frequency sidepeak actually cannot be distinguished as separate features of the spectrum when the drive strength becomes as strong as $\Omega/\omega_c \approx 0.5$. It should be noted that there is noise in all data plotted in Fig. 4.6 due to lack of complete numerical convergence of the methods used, though all results were found to be close to convergence and are quantitatively correct. This lack of fully converged results is not due to a specific limitation of the method but rather because of limited computation time.

When the drive strength is increased to the regime $\Omega/\omega_c \geq 1$ there is no distinguishable phonon sideband and the lower frequency sidepeak becomes the dominant feature of the fluorescence spectra, so this peak alone is plotted in Fig. 4.6b. With increasing drive the height of this peak decreases to a minimum value at $\Omega/\omega_c \approx 3$

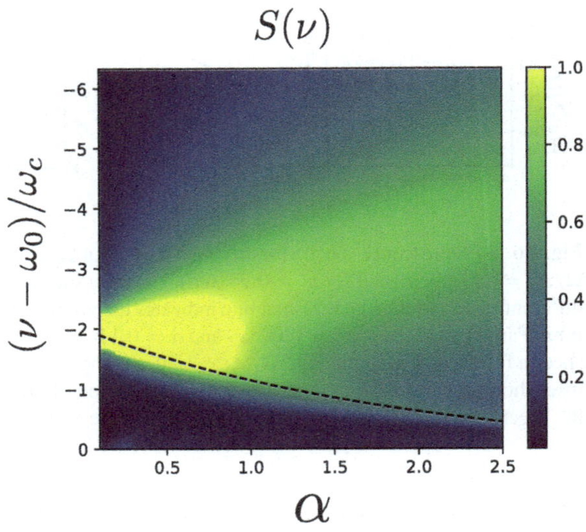

Fig. 4.7 A heat map depicting the negative frequency sidepeak of fluorescence spectra as a function of coupling strength, normalised to have maximum value equal to 1. The black dashed line indicates the position of the peak predicted using a polaron frame. Parameters used are $\gamma = 0.001\omega_c$ and $\Omega = 2\omega_c$

before beginning to increase again. An analysis of the dissipative influence of the phonons using the RSM approximations would result in a description of their effects through a dissipator that causes decay of the exciton into the lower dressed state at a rate $\sim J(\Omega)$, which for SD of this model, Eq. (4.10) has maximum at $\Omega/\omega_c = 3$. In turn this would imply a Lorentzian profile for the lower frequency sidepeak whose width, $w(\Omega)$, and height, $h(\Omega)$, are related by $w(\Omega) \sim h^{-1}(\Omega) \sim J(\Omega)$. The phonon coupling used in Fig. 4.6b is reasonably small, $\alpha = 0.1$, so it is plausible that using the RSM approximations can be justified, despite the relatively long relaxation time of phonons compared to the exciton transition time in this regime. The inverse height, $h^{-1}(\Omega)$, of the numerically calculated sidepeak is plotted as a function of Ω in the inset of Fig. 4.6b and is found to be well described by the curve $J(\Omega)/\pi$, which could confirm this. A quantitative analysis of the width of the peaks as a function of drive is not undertaken here, but qualitatively it can be seen in Fig. 4.6b that $w(\Omega)$ does broaden and then become slightly narrower again the driving is increased through $\Omega = 3\omega_c$. As the drive strength is increased even further the exciton is expected to decouple from the phonons [38], which has been demonstrated in experiment [39]. Future studies will include an exploration of fluorescence spectra in this regime.

So far all results have been obtained in the regime of relatively weak coupling between exciton and phonons. In Fig. 4.7 the behaviour of the lower frequency sidepeak is calculated for a fixed, and large, drive strength $\Omega/\omega_c = 2$ as the coupling strength is increased into the strong coupling regime. At low coupling, $\alpha < 1$, there is a single sidepeak at $\sim -\Omega$ and no phonon sideband. This single peak splits into two in the strong coupling regime $\alpha > 1$. The lower frequency peak in the strong coupling regime has larger height and width than the higher frequency peak and is possibly a phonon sideband, though there is little discussion on fluorescence spectra in this strong driving, strong coupling regime in the literature which could confirm this. The higher frequency peak in Fig. 4.7 can remain interpretable as the low frequency sidepeak of a triplet, however. This is evidenced by the fact that the position of its maximum is well described by the position of such a sidepeak predicted using the renormalised drive strength given by Eq. (4.16), which is shown by the black dashed line in Fig. 4.7.

4.4 Summary

To demonstrate the applicability and flexibility of the methods presented in Chap. 3, the numerical study of three models of open quantum systems has been undertaken. In Sect. 4.1 the well studied problem of the localisation phase transition in the spin-boson model was addressed. The location of the transition was identified through high precision calculations of the dissipative dynamics of the spin. In the case of a spin-$1/2$ the transition point was found to be in agreement with other analytic and numerical studies that have been performed. The capability of the method of handling larger systems was also demonstrated by investigating this transition in the case of a spin-1. A quite different model was studied in Sect. 4.2, that of a pair of spatially separated spins interacting with the same environment. The widely separated timescales present in the dissipative dynamics of the spins make this a very difficult problem to treat with existing methods. Here the problem was solved with relative ease and the dynamics of the effective interaction between the two spins induced by the environment were seen clearly, a result which is otherwise absent from existing literature. Finally, the potential for the use of the methods in Chap. 3 to make experimentally relevant predictions was shown in Sect. 4.3. The behaviour of the fluorescence spectrum of a driven quantum dot was investigated from weak to strong driving and weak to strong exciton-phonon coupling. The phonon sideband in such spectra was found numerically and was shown to be suppressed as the quantum dot is increasingly strongly driven, while the behaviour of the triplet structure in the spectra due to dressed state transitions was found to be intuitive and consistent with analytic results. The regime of both strong drive and strong dissipation was also investigated and again the results obtained were consistent with analytic predictions.

References

1. Leggett AJ, Chakravarty S, Dorsey AT, Fisher MPA, Garg A, Zwerger W (1987) Dynamics of the dissipative two-state system. Rev Mod Phys 59(1):1
2. Magazzù L, Forn-Díaz P, Belyansky R, Orgiazzi J-L, Yurtalan MA, Otto MR, Lupascu A, Wilson CM, Grifoni M (2018) Probing the strongly driven spin-boson model in a superconducting quantum circuit. Nat Commun 9:1403
3. Forn-Díaz P, Romero G, Harmans CJPM, Solano E, Mooij JE (2016) Broken selection rule in the quantum Rabi model. Sci Rep 6:26720

4. Leppäkangas J, Braumüller J, Hauck M, Reiner J-M, Schwenk I, Zanker S, Fritz L, Ustinov AV, Weides M, Marthaler M (2018) Quantum simulation of the spin-boson model with a microwave circuit. Phys Rev A 97(5):052321
5. Ficheux Q, Jezouin S, Leghtas Z, Huard B (2018) Dynamics of a qubit while simultaneously monitoring its relaxation and dephasing. Nat Commun 9:1926
6. Florens S, Venturelli D, Narayanan R (2010) Quantum Quenching, Annealing and Computation. Springer, Berlin, p 145
7. Le Hur K (2010) Quantum phase transitions in spin-boson systems: dissipation and light phenomena. In Understanding Quantum Phase Transitions. CRC Press, US
8. Vojta M, Tong N-H, Bulla R (2005) Quantum phase transitions in the sub-ohmic spin-boson model: failure of the quantum-classical mapping. Phys Rev Lett 94(7):070604
9. Chin AW, Prior J, Huelga SF, Plenio MB (2011) Generalized polaron ansatz for the ground state of the sub-ohmic spin-boson model: an analytic theory of the localization transition. Phys Rev Lett 107(16):160601
10. Duan C, Tang Z, Cao J, Wu J (2017) Zero-temperature localization in a sub-ohmic spin-boson model investigated by an extended hierarchy equation of motion. Phys Rev B 95(21):214308
11. Amann A (1991) Ground states of a spin-boson model. Ann Phys 208(2):414
12. Wang H, Thoss M (2008) From coherent motion to localization: dynamics of the spin-boson model at zero temperature. New J Phys 10(11):115005
13. Strathearn A, Kirton P, Kilda D, Keeling J, Lovett BW (2018) Efficient non-Markovian quantum dynamics using time-evolving matrix product operators. Nat Commun 9:3322
14. Bulla R, Tong N-H, Vojta M (2003) Numerical renormalization group for bosonic systems and application to the sub-ohmic spin-boson model. Phys Rev Lett 91:170601
15. Weiss U (2012) Quantum Dissipative Systems. World Scientific
16. Orth PP, Roosen D, Hofstetter W, Le Hur K (2010) Dynamics, synchronization, and quantum phase transitions of two dissipative spins. Phys Rev B 82:144423
17. Winter A, Rieger H (2014) Quantum phase transition and correlations in the multi-spin-boson model. Phys Rev B 90:224401
18. McCutcheon DPS, Nazir A (2010) Quantum dot Rabi rotations beyond the weak exciton-phonon coupling regime. New J Phys 12(11):113042
19. Andersson G, Suri B, Guo L, Aref T, Delsing P (2018) Nonexponential decay of a giant artificial atom. arXiv:1812.01302
20. Facchi P, Pascazio S, Pepe FV, Pomarico D (2018) Correlated photon emission by two excited atoms in a waveguide. Phys Rev A 98(6):063823
21. Mahan GD (2000) Many Particle Physics, 3rd edn. Springer, US
22. McCutcheon DPS, Nazir A (2011) Coherent and incoherent dynamics in excitonic energy transfer: correlated fluctuations and off-resonance effects. Phys Rev B 83:165101
23. Nalbach P, Eckel J, Thorwart M (2010) Quantum coherent biomolecular energy transfer with spatially correlated fluctuations. New J Phys 12(6):065043
24. Hanschke L, Fischer KA, Appel S, Lukin D, Wierzbowski J, Sun S, Trivedi R, Vučković J, Finley J, Müller K (2018) Quantum dot single-photon sources with ultra-low multi-photon probability. npj Quantum Inf 4:43
25. Senellart P, Solomon G, White A (2017) High-performance semiconductor quantum-dot single-photon sources. Nat Nanotechnol 12(11):1026
26. Unitt DC, Bennett AJ, Atkinson P, Cooper K, See P, Gevaux D, Ward MB, Stevenson RM, Ritchie DA, Shields AJ (2005) Quantum dots as single-photon sources for quantum information processing. J Opt B: Quantum Semiclass Opt 7(7):S129
27. Loss D, DiVincenzo DP (1998) Quantum computation with quantum dots. Phys Rev A 57(1):120
28. Kloeffel C, Loss D (2013) Prospects for spin-based quantum computing in quantum dots. Ann Rev Condens Matter Phys 4(1):51
29. Lovett BW, Reina JH, Nazir A, Briggs AI (2003) Optical schemes for quantum computation in quantum dot molecules. Phys Rev B 68(20):205319

30. Bera D, Qian L, Tseng T-K, Holloway PH (2010) Quantum dots and their multimodal applications: a review. Materials 3(4):2260
31. Godden TM, Ramsay AJ, Boyle SJ, Gauger EM, Nazir A, Lovett B, Fox M, Skolnick M (2011) Intensity damping of Rabi-oscillations and renormalization of the Rabi frequency in InGaAs/GaAs quantum dots. In CLEO:2011 - Laser Applications to Photonic Applications. OSA
32. Peter E, Hours J, Senellart P, Vasanelli A, Cavanna A, Bloch J, Gérard JM (2004) Phonon sidebands in exciton and biexciton emission from single GaAs quantum dots. Phys Rev B 69(4)
33. Roy C, Hughes S (2012) Polaron master equation theory of the quantum-dot Mollow triplet in a semiconductor cavity-QED system. Phys Rev B 85(11):115309
34. McCutcheon DPS (2016) Optical signatures of non-Markovian behavior in open quantum systems. Phys Rev A 93(2):022119
35. Iles-Smith J, McCutcheon DPS, Nazir A, Mørk J (2017) Phonon scattering inhibits simultaneous near-unity efficiency and indistinguishability in semiconductor single-photon sources. Nat Photonics 11(8):521
36. Tighineanu P, Dreeßen CL, Flindt C, Lodahl P, Sørensen AS (2018) Phonon decoherence of quantum dots in photonic structures: broadening of the zero-phonon line and the role of dimensionality. Phys Rev Lett 120(25):257401
37. Nazir A, McCutcheon DPS (2016) Modelling exciton-phonon interactions in optically driven quantum dots. J Phys Condens Matter 28(10):103002
38. Vagov A, Croitoru MD, Axt VM, Kuhn T, Peeters FM (2007) Nonmonotonic field dependence of damping and reappearance of Rabi oscillations in quantum dots. Phys Rev Lett 98(22):227403
39. Kaldewey T, Lüker S, Kuhlmann AV, Valentin SR, Chauveau J-M, Ludwig A, Wieck AD, Reiter DE, Kuhn T, Warburton RJ (2017) Demonstrating the decoupling regime of the electron-phonon interaction in a quantum dot using chirped optical excitation. Phys Rev B 95(24):241306

Chapter 5
Conclusion

The development of a theory of open quantum systems has had great impact on how the microscopic world is understood and realising how quantum mechanical behaviour can exploited in applications. The set up of the problem of calculating observables and making predictions about an open quantum system is formally rather simple, as shown in Sects. 2.1.1.1 and 2.1.1.2. The mathematical concepts required to write down the exact expressions one must evaluate to solve arbitrary problems are basic; using elementary tools of linear algebra, calculus, and probability theory, the closed form equations that describe arbitrary open quantum systems can be derived. All such equations take the form of time ordered influence functionals. Despite the relative ease with which problems in open quantum systems can be set up, the formal nature of the theory means that solutions generally must be found on a problem-by-problem basis, requiring further mathematical machinery and rather specific physical insight. Only for one class of systems do there exist theoretical tools that can be applied universally to solve problems easily. These are Markovian open systems. In Sect. 2.1.2.2 it was shown how approximations, underpinned by the assumptions of Markovianity and weak coupling, are made on the exact equations describing open quantum systems. This reduces the problem of evaluating time ordered influence functionals to a simple exercise in linear algebra.

Although Markovian systems are most abundant the property of Markovianity is itself only ever approximate [1], in the same way that any realistic physical system can only ever be considered as approximately closed. The ability to probe and control open quantum systems ever more precisely therefore demands an understanding of their non-Markovian behaviour. The general physical principles of non-Markovian systems are not well established, a result of the lack of techniques that can be applied universally to investigate such systems. The contents of Sect. 2.1.3 provided a review of methods that can be used to investigate special cases of non-Markovian systems. These include cases in which the time ordered influence functionals that describe an

A. Strathearn, *Modelling Non-Markovian Quantum Systems Using Tensor Networks*, Springer Theses, https://doi.org/10.1007/978-3-030-54975-6_5

open system can be evaluated exactly, either analytically or numerically, unitarily transforming the open system to make it appear Markovian, and explicitly modelling the parts of the environment responsible for non-Markovianity. Exact methods do exist for probing the non-Markovian nature of open systems, though almost all are tailored specifically for systems whose environment is a harmonic bath, the type of environment discussed in Sect. 2.1.1.3, and are too computationally expensive to be applied to systems much larger than a two-level system.

The methods for exactly calculating observables of non-Markovian systems introduced in Chap. 3, the main result of this thesis, have greater scope than other techniques currently available in the field. It was shown analytically in Sect. 3.1 that application of the time ordering operator to any expression can be carried out by contracting a tensor network. This general result is clearly relevant to the field of open quantum systems, but has potential application anywhere a mathematical ordering operation is encountered. Tensor networks, as discussed in Sect. 2.2, have found widespread use in solving problems in mathematical sciences. Indeed, many of the best existing methods developed to be applicable to specific non-Markovian open systems involve representation of a problem as a tensor network. The connection between generic influence functionals and tensor networks shown in Sect. 3.2, however, provides a novel and general link between the field of open quantum systems and that of tensor networks. Given the vast and fast growing literature on tensor networks, this link potentially provides a completely new theoretical means by which open quantum systems can be described.

In Sect. 3.3 it was shown how the tensor network representing a particular form of time ordered influence functional can be cast in the form of contractions between matrix product states and operators. The type of influence functional in question could represent exactly that of a system coupled to a harmonic bath, though it was also shown in Sect. 3.2 how nonperturbative approximations can be made to put arbitrary influence functionals in the form that admits a matrix product representation. These approximations are a generalisation of the type used when seeking a Markovian description of an open system, making the overall method well-aligned with established open systems theory. It was further shown in Sect. 3.3 how the standard discretised path integral description of a system coupled to a harmonic bath can be recovered from its corresponding tensor network representation. The standard techniques used to efficiently contract tensor networks that have matrix product form was discussed in Sect. 2.2.3, and in Sect. 3.6 these were applied to the tensor network representation of a time ordered influence functional to show that the method is not just theoretically appealing but is indeed of practical use. An observable of a two-level system coupled to a harmonic bath was calculated and the efficiency of the calculation was compared to the efficiency of the same calculation performed using the standard path integral method. It was found that exploiting the tensor network structure of the problem improved the efficiency by many orders of magnitude.

The calculations performed in Sect. 3.6 provided a proof-of-principle for the methods presented in Chap. 3. The contents of Chap. 4 are further applications of the method to more interesting open quantum systems, and provide evidence of the generality and range of applicability of the method. First, in Sect. 4.1, it was shown how a dissipative phase transition could be identified in the spin-boson model. The phase transition in question is a localisation transition, making it intrinsically non-Markovian and nontrivial in terms of dynamics of the system. There are both analytical [2, 3] and numerical [4], results surrounding the nature of this phase transition when the spin is a spin-$1/2$, and dynamics calculated here produced results that were in agreement with the literature. The efficiency of using tensor networks to solve the problem also allowed the consideration of the spin-1 phase transition, a problem that currently lies on the boundary of what is solvable using other techniques. Calculations for this problem were performed in a straightforward manner, producing results that were in agreement with some literature and not others, showing that the methods developed in this thesis have the potential to resolve outstanding problems in the field. Not shown in this section are calculations that were performed which show the phase transition for up to spin-$9/2$, a ten-level problem which no other methods are capable of addressing. Work on this problem is ongoing and will be returned to in the future.

The next system to be studied was that of two spins interacting with a common environment, the results of which were presented in Sect. 4.2. This problem is difficult to tackle with existing methods due to the spins having finite separation in space. The dynamics of the spins are determined both by coupling to their immediate environment and by the effective interaction between them induced by environmental excitations propagating between the two. This environment mediated interaction between the spins is a highly non-Markovian phenomena, since it depends solely on how the environment is perturbed by the spins and how this perturbation affects the spins at a later time. There are no results in the literature relating to the parameter regimes that were investigated for this system and similarly no existing techniques suitable for application to this problem, yet the use of tensor networks here allowed easy solution to the problem. The effects of changing the spatial dimension of the environment and separation of the spins were investigated, providing physically intuitive results that confirmed the system behaves as expected. Although the results of Sect. 4.2 simply confirm intuition about the problem of two spins interacting with a common environment, they would not otherwise have been attainable without the use of the methods in Chap. 3. Moreover, this problem is very different in nature to the problem of localisation in the spin-boson model, highlighting how the effective use of tensor networks has not relied on any particular feature of the models involved.

As a more practical application of the method, the emission spectra of a driven quantum dot were calculated in Sect. 4.3 for various parameters. The dynamical behaviour of vibrational degrees of freedom that couple to the driven exciton transition in the quantum dot result in a phonon sideband in emission. The phonon sideband is reasonably well understood theoretically and is readily observable in experiment. The limit of weak driving of the dot is accessible analytically through perturbation theory and the limit in the strong driving limit the exciton decouples from the phonons. However, the intermediate regime, where phonon relaxation and exciton transitions occur on a similar timescale, is relatively unexplored. Emission spectra across a large range of drive strengths were calculated using tensor networks, with the weak driving limit appearing as expected and the intermediate regime displaying intuitive crossover behaviour between strong and weak drive. These results demonstrate that the methods of Chap. 3 are applicable to experimentally relevant systems and can be used to calculate nontrivial experimentally accessible quantities that display non-Markovian features.

Although the physical systems that were investigated in Chap. 4 are rather simple, they exhibit a diverse range of non-Markovian phenomena and yet the use of the tensor networks methods of Chap. 3 made their numerical solution straightforward. No other methods to date allow for such unified treatment with little regard to details of the models. Both single and multitime observables were calculated within the same methodological framework and many of the calculations performed can easily be carried out on the order of minutes on a laptop computer. Although it is not yet clear how the truncation of singular values that enables such efficiency can be justified on physical grounds, numerical evidence has shown that this approach is indeed highly effective. Thus, as it stands, the only limitation of the method appears to be that of the size of the individual tensors that must be constructed prior to performing any contractions, which grows exponentially with the size of the system being considered. Even without knowledge of the theory behind the methods presented in Chap. 3 it is relatively straightforward to use the simulation technique as a blackbox, both for benchmarking and performing exploratory studies of non-Markovian open quantum systems. A Python implementation of the technique that can be used in such a blackbox manner can be accessed through the link provided through Ref. [5] (see also the 'Digital Outputs Access Statement' at the start of this thesis).

The identification of the time ordered influence functionals that are ubiquitous in the theory of open quantum systems with tensor networks will surely open many new doors in the field. The applications given in this thesis simply represent the all important first step that establishes basic validity and practicality. Already the potential widespread range of use and generality of the methods presented in this thesis are being explored. For instance, the method has found application in quantum thermometry [6] and has also been used to investigate systems composed of continuous degrees freedom [7] and those with a highly structured environments [8]. In the latter of these works it was also shown how the method can even be used to calculate observables of the environment, not just of the open system. Flexibility of the method means that open systems with time dependent Hamiltonian can also be considered and is the subject of ongoing work. In terms of development of the method, it has already been shown that different tensor network contraction schemes can improve efficiency even further [9]. Although all systems studied in Chap. 3 involve only bosonic environments, the theory behind the method is, in principle, applicable to arbitrary environments, and development of a practical implementation that applies to fermionic environments is now being explored. Furthermore, the connection made between the tensor network and discretised path integral representations of open quantum systems in Sect. 3.3 means that advanced path integral methods [10–12], such as the calculation of environmental observables using an influence functional [13], can now be implemented using tensor networks. Another potential avenue of development is the extension of the method to the case of a many body non-Markovian quantum system by using established tensor network representations spatially extended states [14]. Basic comparisons of the method to existing master equation techniques were made in Ref. [8], but a comprehensive comparison with alternative state-of-the-art methods in the field of open quantum systems [15–17] will be the subject of future work.

References

1. Figueroa-Romero P, Modi K, Pollock FA (2019) Almost Markovian processes from closed dynamics. Quantum 3:136
2. Leggett AJ, Chakravarty S, Dorsey AT, Fisher MPA, Garg A, Zwerger W (1987) Dynamics of the dissipative two-state system. Rev Mod Phys 59(1):1
3. Le Hur K (2010) Quantum phase transitions in spin-boson systems: Dissipation and light phenomena. In Understanding Quantum Phase Transitions. CRC Press, US
4. Wang H, Thoss M (2008) From coherent motion to localization: dynamics of the spin-boson model at zero temperature. New J Phys 10(11):115005
5. Strathearn A, Kilda D (2018) Python implementation of TEMPO. https://doi.org/10.5281/zenodo.1322407
6. Jrgensen MR, Potts PP, Paris MGA, Brask JB (2020) Tight bound on finite-resolution quantum thermometry at low temperatures. arXiv:2001.04096
7. Minoguchi Y, Kirton P, Rabl P (2019) Environment-Induced Rabi Oscillations in the Optomechanical Boson-Boson Model. arXiv:1904.02164
8. Gribben D, Strathearn A, Iles-Smith J, Kilda D, Nazir A, Lovett BW, Kirton P (2020) Exact quantum dynamics in structured environments. Phys Rev Res 2:013265
9. Jørgensen MR, Pollock FA (2019) Exploiting the causal tensor network structure of quantum processes to efficiently simulate non-Markovian path integrals. Phys Rev Lett 123:240602
10. Sim E, Makri N (1997) Filtered propagator functional for iterative dynamics of quantum dissipative systems. Comput Phys Commun 99(2–3):335
11. Makri N (2017) Iterative blip-summed path integral for quantum dynamics in strongly dissipative environments. J Chem Phys 146(13):134101
12. Segal D, Millis AJ, Reichman DR (2010) Numerically exact path-integral simulation of nonequilibrium quantum transport and dissipation. Phys Rev B 82(20):205323
13. Kilgour M, Agarwalla BK, Segal D (2019) Path-integral methodology and simulations of quantum thermal transport: full counting statistics approach. J Chem Phys 150(8):084111
14. Suzuki S, Oshiyama H, Shibata N (2019) Quantum annealing of pure and random Ising chains coupled to a Bosonic environment. J Phys Soc Jpn 88(6):061003
15. Suess D, Eisfeld A, Strunz WT (2014) Hierarchy of stochastic pure states for open quantum system dynamics. Phys Rev Lett 113(15):150403
16. Prior J, Chin AW, Huelga SF, Plenio MB (2010) Efficient simulation of strong system-environment interactions. Phys Rev Lett 105(5):050404
17. Chin AW, Prior J, Rosenbach R, Caycedo-Soler F, Huelga SF, Plenio MB (2013) The role of non-equilibrium vibrational structures in electronic coherence and recoherence in pigment-protein complexes. Nat Phys 9(2):113

The manufacturer's authorised representative in the EU is Springer Nature Customer Service Centre GmbH, Europaplatz 3, 69115 Heidelberg, Germany. If you have any concerns regarding our products, please contact ProductSafety@springernature.com

Printed and bound by CPI Group (UK) Ltd, Croydon, CR0 4YY
30/09/2026
02225515-0001